MAMMALS IN COLOUR

MAMMALS
IN COLOUR

by

LEIF LYNEBORG

Illustrations by

HENNING ANTHON

Translated and adapted by

GWYNNE VEVERS & WINWOOD READE

BLANDFORD PRESS
LONDON

First English edition 1971

167 High Holborn, London WC1V 6PH

ISBN 07137 0548 5

Colour sheets printed in Denmark by F. E. Bording A/S
Set in Photon Times 9 pt. by
Richard Clay (The Chaucer Press), Ltd., Bungay, Suffolk
and printed in Great Britain by
Fletcher & Son, Ltd., Norwich, Norfolk

CONTENTS

Contents

PREFACE

Mammals in Colour is designed as a handbook for the identification of the mammals of Britain and Europe. Species are illustrated in colour: the majority of the plates show each animal in as natural a position as possible. When it is necessary to illustrate differences between the sexes, seasonal changes in coat and so on, there is more than one drawing for the species. There are many detailed illustrations of features which help in the identification of closely related species and also provide information about the habits of various animals: these include drawings of skulls, teeth, antlers, tracks and so on.

The second section of the book contains descriptive text. There is a short introduction to each of the systematic groups and an account of each species shown in the colour plates, as well as of a few others. Under each species the text is divided into four parts:

Identification: This gives measurements, weights and any other significant characters which help in identification. Body-length denotes the distance from the tip of the muzzle to the base of the tail. Tail-length represents the distance from the base to the tip of the tail. Shoulder-height, given for certain of the larger mammals, is measured from the ground to the highest point of the back near the shoulder. In some of the smaller mammals the length of the hind-foot is significant and this is measured from the back of the heel to the tip of the longest toe. In the bats, measurements are also given for the length of the ear, the tragus, the forearm and the wing-span. Measurements and weights are given first in metric units as these are the most accurate; the conversions into inches, feet, ounces and so on should only be regarded as a rough guide. The figures given apply only to adults as measurements and weights naturally vary considerably according to age and condition.

Distribution: This section describes the geographical range of the species. This can only be a broad outline for it should be remembered that many mammals have a somewhat sporadic distribution.

Habitat: This gives the type of environment in which the species normally occurs.

Habits: This section gives general biological information. It includes the type of home, periods of activity and rest, senses, breeding data, populations fluctuations, voice, life-span, enemies and so on.

The original text is by Leif Lyneborg and the colour plates and line drawings are by Henning Anthon. In the preparation of both text and plates valuable assistance has been given by Preben Bang and Poul Valentin Jensen. The distribution maps were drawn by Arne Gaarn Bak.

INTRODUCTION TO MAMMALS

The mammals are a group with relatively few species. At the present time it is estimated that there are about 4,500 living species distributed throughout the world. To set the mammals in perspective: there are about 8,600 living species of bird, about 23,000 species of fish and upwards of a million species of insect. There is no doubt that the mammal fauna of the present time is considerably poorer than that which lived in the late Tertiary Period—a few million years ago. A great number of species have become extinct. In fact, only a third of all the genera have living representatives. Of the 32 known orders of mammals only the rodents and the artiodactyls have not decreased in number of species since the Tertiary Period.

The mammal fauna of Europe has many fewer species than that of the other continents. During the last Ice Age the climate over large areas of Europe was unsuitable for most mammals. In post-glacial times different climatic conditions brought about gradual but radical changes in the natural vegetation which resulted in changes in the composition of the mammalian fauna. Evidence of these changes is found in the form of fossil remains—in bogs, gravel pits, quarries and similar places.

All mammals have a number of characteristics in common. They breathe with lungs, the heart has 4 chambers and they have a complicated arrangement of 3 bones in the middle ear. In addition, they maintain a constant high body temperature; a few hibernating species show only a temporary decrease in body temperature.

Hair is a special structure not found in any other vertebrates. The hairs lie in deep pits in the skin, known as hair follicles, and are formed of dead horny (keratinized) epidermal cells. They are shed at various moulting periods. Often there are hairs of different types in the same mammal. Thus, there is usually an under layer of woolly hairs and an outer layer of guard hairs; the latter gives the coat its colour. The whiskers or *vibrissae*, which occur mainly on the muzzle, are long hairs and as their follicles are richly supplied with nerves they function as sensitive tactile organs. The mammary glands which produce milk, are also unique to the mammals in the animal kingdom.

The teeth of mammals require some description: they can be differentiated into incisors, canines and cheek teeth. The original number of teeth is probably 44, with 3 incisors, 1 canine and 7 cheek teeth in each half-jaw. The 4 front cheek teeth are the premolars, the rear 3 the molars. The original dentition is found in the order Insectivora, but in other orders it is much reduced and specialized. As a rule the incisors are chisel-shaped, the canines conical, while the cheek teeth may have points (cusps) or projecting ridges of enamel. Young mammals have a milk dentition which is later replaced by the permanent teeth; the molars, however, have no precursors in the milk dentition.

There is little doubt that the mammals originated as land animals. This means that mammal groups such as the seals and whales, which are both highly specialized for aquatic life although not closely related to one another, have descended from forms which lived on land. Mammals occur in all climates from the arctic wastes, through the temperate forest areas to the subtropics and tropics. A remarkable capacity for adapting themselves to different living conditions has produced enormous variety in their structure and habits.

The living mammals are classified in 19 orders, of which 9 are represented in Europe. A key to these 9 orders is given below. The orders are divided into families, each with a Latin name ending in -idae, and each family has one or more genera with a Latin name beginning with a capital letter, e.g. *Sorex*. Each genus contains one or more species, with Latin names beginning with a lower-case letter, e.g. *araneus*. Finally a species may be divided into a number of subspecies or races.

The scientific nomenclature used in this book follows the *Checklist of Palaearctic and Indian mammals 1758 to 1946*, by J. R. Ellerman and T. C. S. Morrison-Scott (London, British Museum, 1951).

Key to the mammalian orders of Europe

Page

1. Only one pair of visible limbs ....Whales (Cetacea) 230
   Two pairs of visible limbs ....see 2

2. Both pairs of limbs in the form of flippers, the front ones directed outwards, the rear ones backwards ....Seals (Pinnipedia) 205
   Limbs not in the form of flippers ....see 3

3. Front limbs developed as wings, with a wing-membrane stretched between the arms and the much elongated fingers ....Bats (Chiroptera) 115
   Front limbs without a wing-membrane ....see 4

4. Muzzle elongated, forming a long snout ....Insectivores (Insectivora) 101
   Muzzle not elongated..see 5

5. Feet with hooves. Head often with antlers or horns ....Artiodactyls (Artiodactyla) 212
   Feet without hooves. Head always without antlers or horns...........see 6

6. Upper and lower jaw each with 2 large, chisel-shaped incisors. A toothless gap between incisors and cheek teeth. No canines ....see 7
   At least 4 uniform incisors in both upper and lower jaws. A powerful, conical canine immediately behind the incisors ....see 8

Page

7. Four toes on each hind-foot. In the upper jaw 2 very small incisors behind the large incisors ....Lagomorphs (Lagomorpha) 139
   Five toes on each hind-foot. No small incisors behind the large incisors of the upper jaw ....Rodents (Rodentia) 144

8. Toes and fingers elongated. 'Hands' on fore- and hind-limbs ......Primates 228
   Toes and fingers short. 'Paws' on fore- and hind-limbs ....Carnivores (Carnivora) 180

THE COLOUR ILLUSTRATIONS
pp. 13–100

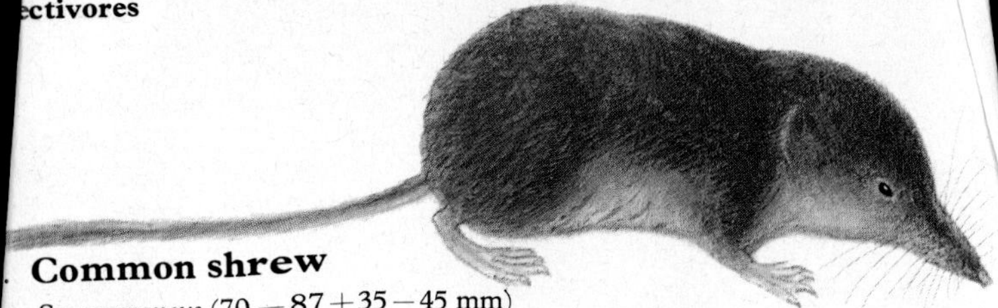

. **Common shrew**

Sorex araneus (70 — 87 + 35 — 45 mm)

a. skull

b. teeth

3

4

4. **Pygmy shrew**

Sorex minutus (43 — 60 + 31 — 46 mm)

a. teeth

5

5. **Laxmann's shrew**

Sorex caecutiens (44 — 67 + 31 — 44 mm)

a. teeth

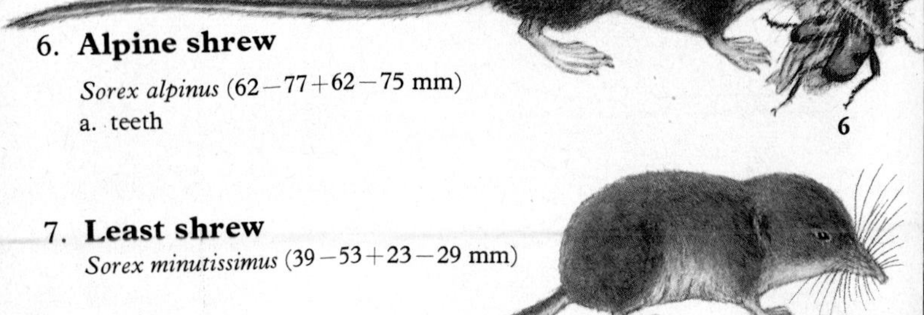

6. **Alpine shrew**

Sorex alpinus (62 — 77 + 62 — 75 mm)

a. teeth

6

7. **Least shrew**

Sorex minutissimus (39 — 53 + 23 — 29 mm)

7

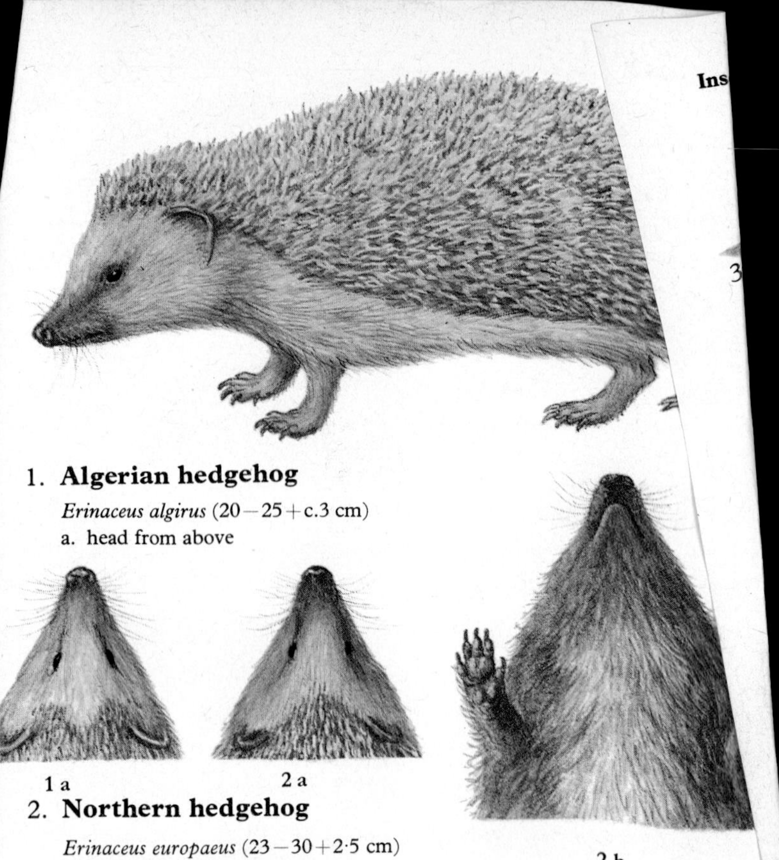

1. **Algerian hedgehog**

   Erinaceus algirus (20 – 25 + c.3 cm)
   a. head from above

2. **Northern hedgehog**

   Erinaceus europaeus (23 – 30 + 2·5 cm)
   a. head from above
   b. eastern form, ventral view

8

8. **Water shrew**

Neomys fodiens (72 – 96 + 47 – 77 mm)

a. left hind-foot
b. swimming, hunting minnows
c. skull

8 a

8 b

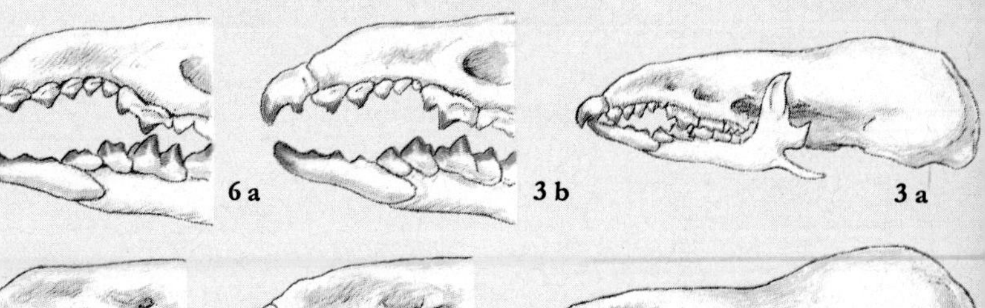

6 a 3 b 3 a

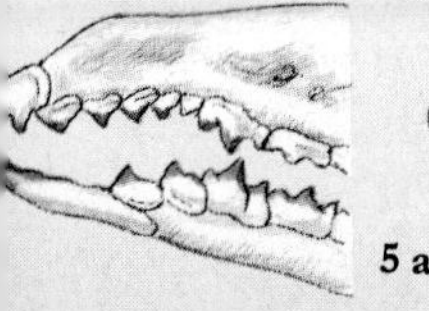

5 a

4 a

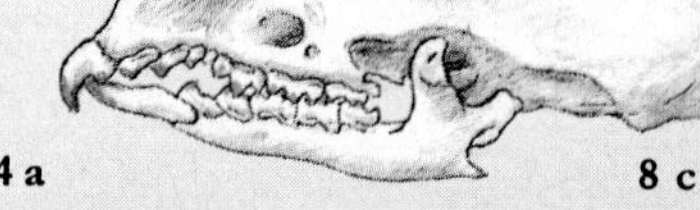

8 c

9. Pygmy white-toothed shrew

Suncus etruscus (34 – 44 + 24 – 29 mm)

9

10

10. Bicolour white-toothed shrew

Crocidura leucodon (64 – 87 + 28 – 39 mm), adult with young in 'caravan' a. teeth

11. Lesser white-toothed shrew

Crocidura suaveolens (53 – 82 + 24 – 44 m

11

12

12. Common white-toothed shrew

Crocidura russula (64 – 95 + 40 – 50 mm)
a. skull b. teeth

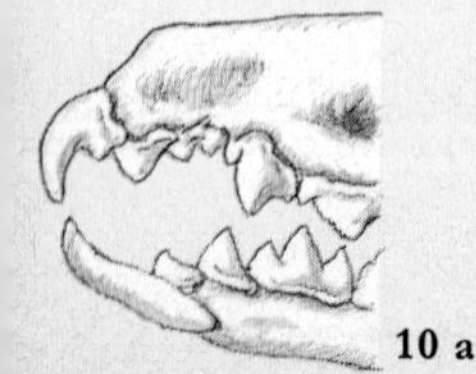

10 a

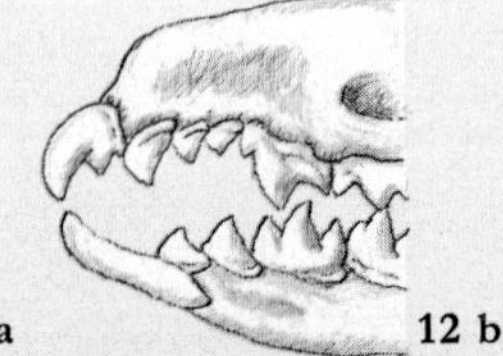

12 b

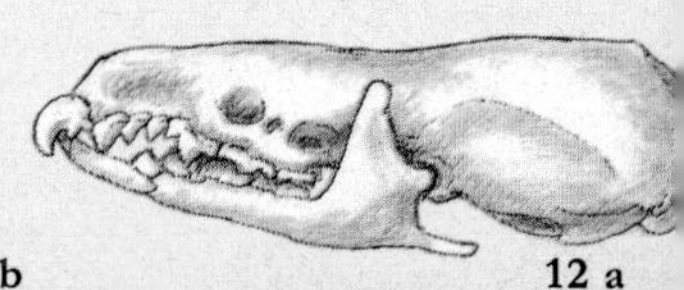

12 a

13

13 a

13. Pyrenean desman

Galemys pyrenaicus (110 – 135 + 130 – 155 mm)
a. left hind-foot, from below

14

14. Northern mole

Talpa europaea (115 – 150 + 20 – 34 mm)
a. right fore-front, front view
b. skull

14 a

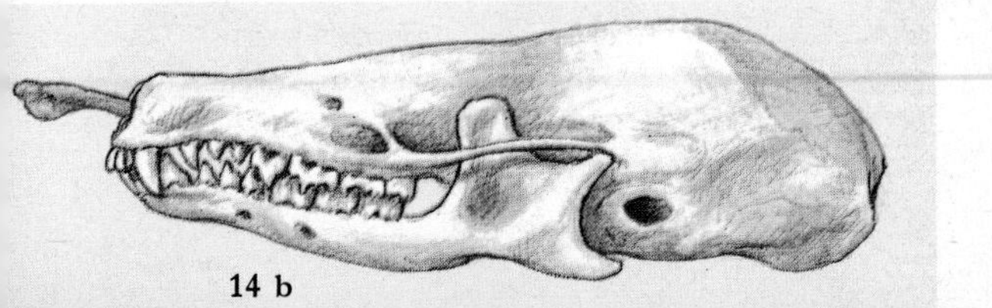

14 b

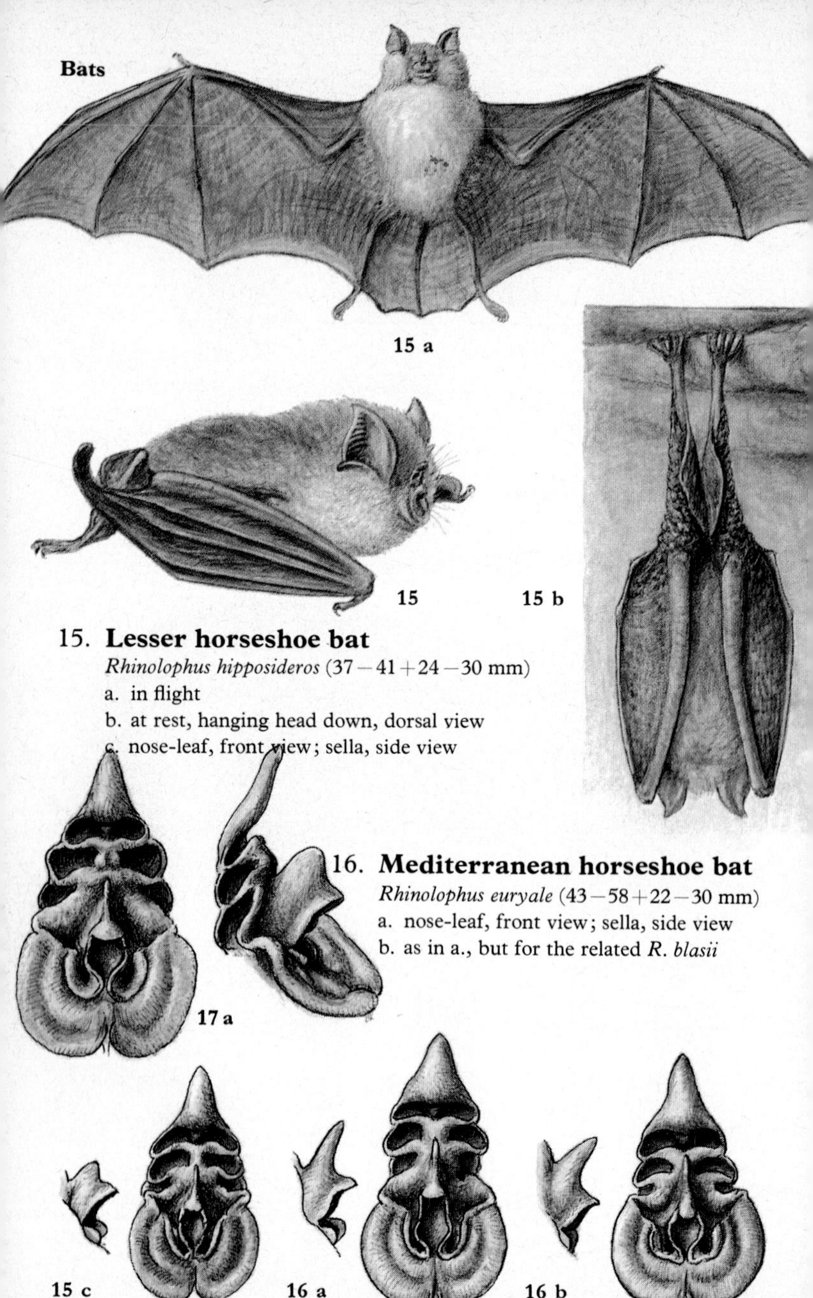

15. **Lesser horseshoe bat**
Rhinolophus hipposideros (37 – 41 + 24 – 30 mm)
a. in flight
b. at rest, hanging head down, dorsal view
c. nose-leaf, front view; sella, side view

16. **Mediterranean horseshoe bat**
Rhinolophus euryale (43 – 58 + 22 – 30 mm)
a. nose-leaf, front view; sella, side view
b. as in a., but for the related *R. blasii*

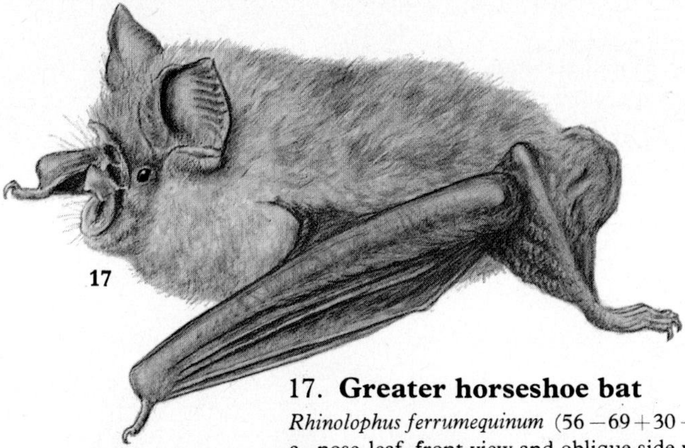

17

17. **Greater horseshoe bat**

Rhinolophus ferrumequinum (56 – 69 + 30 – 43 mm)
a. nose-leaf, front view and oblique side view

18 a

18. **Long-eared bat**

Plecotus auritus
(41 – 51 + 34 – 50 mm)
a. in flight
b. at rest, hanging head down

18 b

18

19. **Schreibers' bat**
Miniopterus schreibersii (52–60+50–60 mm)
a. in flight

20. **Barbastelle**
Barbastella barbastellus (44–58+41–54 mm)
a. head, front view

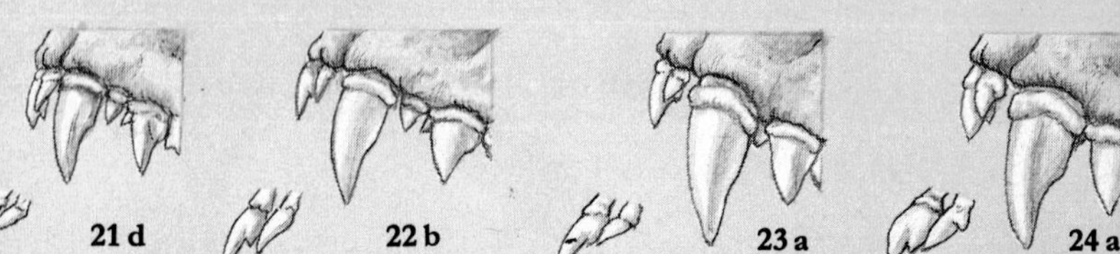

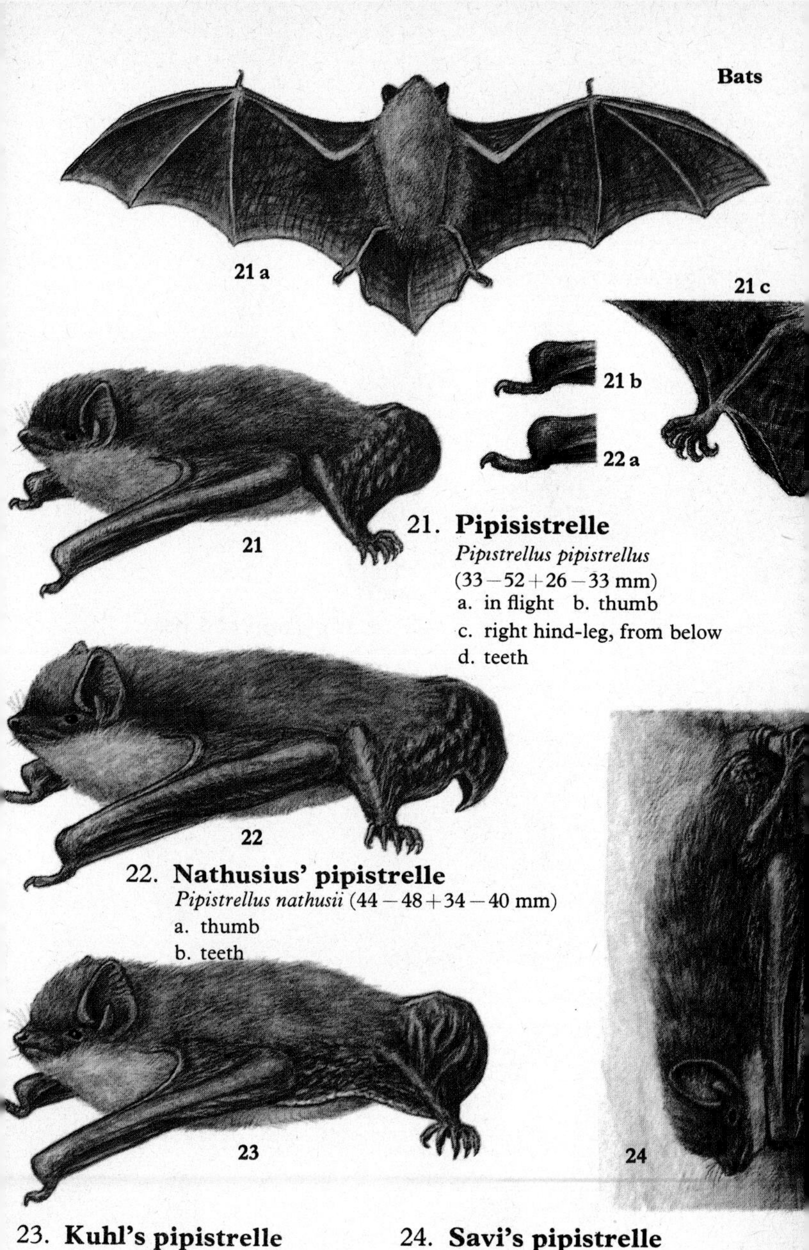

21. **Pipisistrelle**
Pipistrellus pipistrellus
(33 – 52 + 26 – 33 mm)
a. in flight b. thumb
c. right hind-leg, from below
d. teeth

22. **Nathusius' pipistrelle**
Pipistrellus nathusii (44 – 48 + 34 – 40 mm)
a. thumb
b. teeth

23. **Kuhl's pipistrelle**
Pipistrellus kuhli (40 – 47 + 30 – 40 mm)
a. teeth

24. **Savi's pipistrelle**
Pipistrellus savii (43 – 48 + 34 – 39 mm)
a. teeth

25. **Daubenton's or Water bat**
Myotis daubentoni (41 – 51 + 30 – 39 mm)
a. right ear, and right hind-leg and tail, seen from below

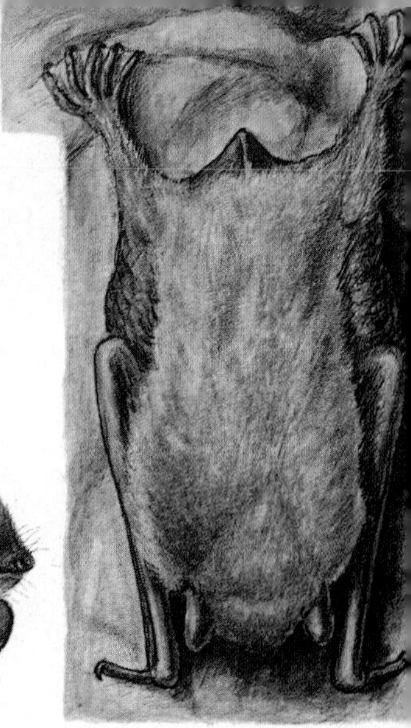

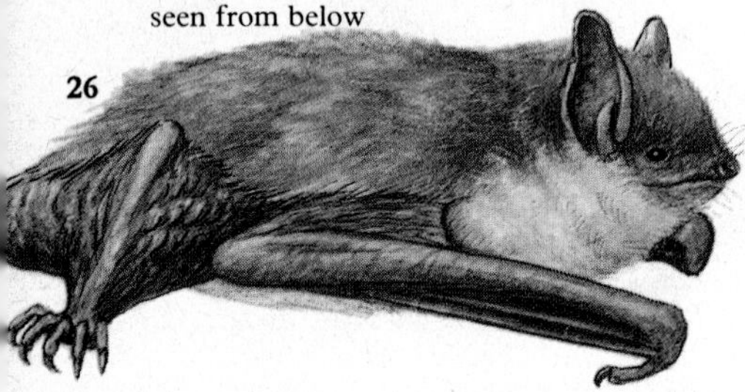

26. **Pond bat**
Myotis dasycneme
(57 – 61 + 46 – 51 mm)
a. as in 25a

27. **Long-fingered bat**
Myotis capaccinii (47 – 53 + 35 – 38 mm)
a. as in 25a

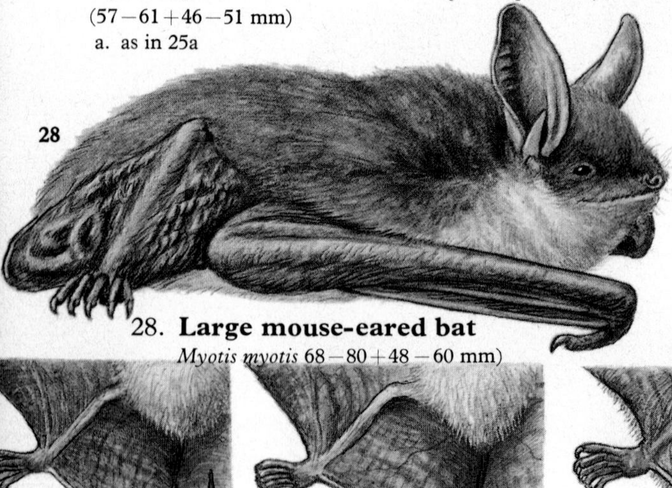

28. **Large mouse-eared bat**
Myotis myotis 68 – 80 + 48 – 60 mm)

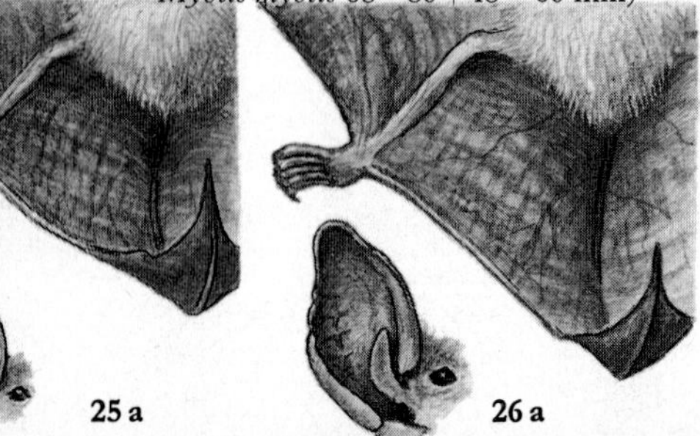

25 a

26 a

27 a

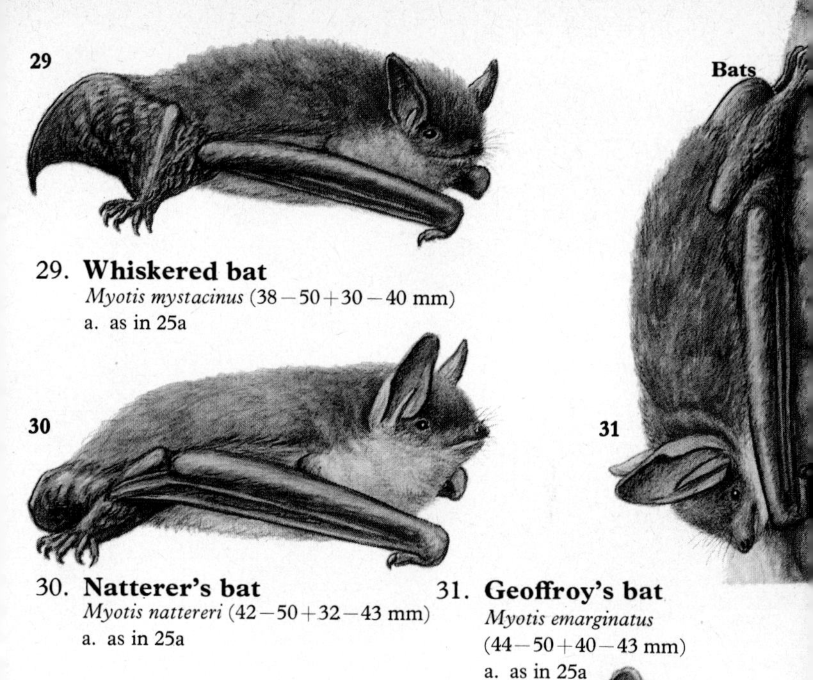

29. **Whiskered bat**
Myotis mystacinus (38 – 50 + 30 – 40 mm)
a. as in 25a

30. **Natterer's bat**
Myotis nattereri (42 – 50 + 32 – 43 mm)
a. as in 25a

31. **Geoffroy's bat**
Myotis emarginatus
(44 – 50 + 40 – 43 mm)
a. as in 25a

32. **Bechstein's bat**
Myotis bechsteini (46 – 53 + 34 – 44 mm)

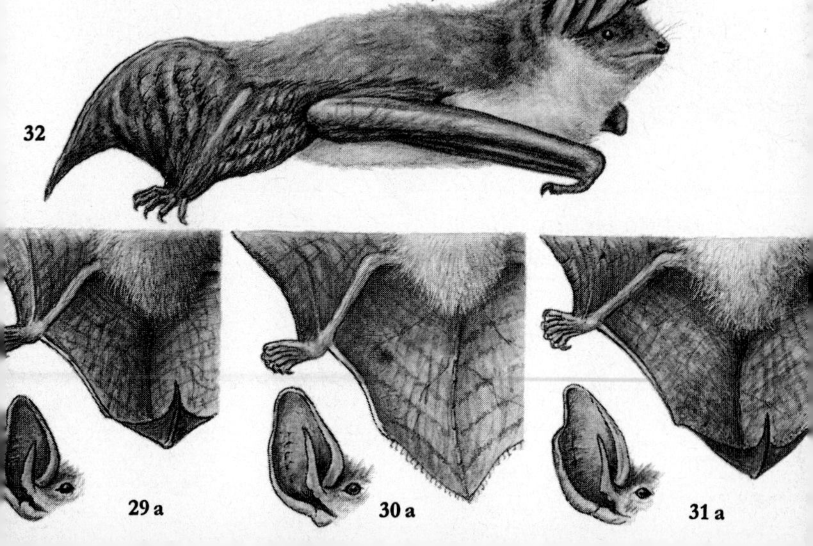

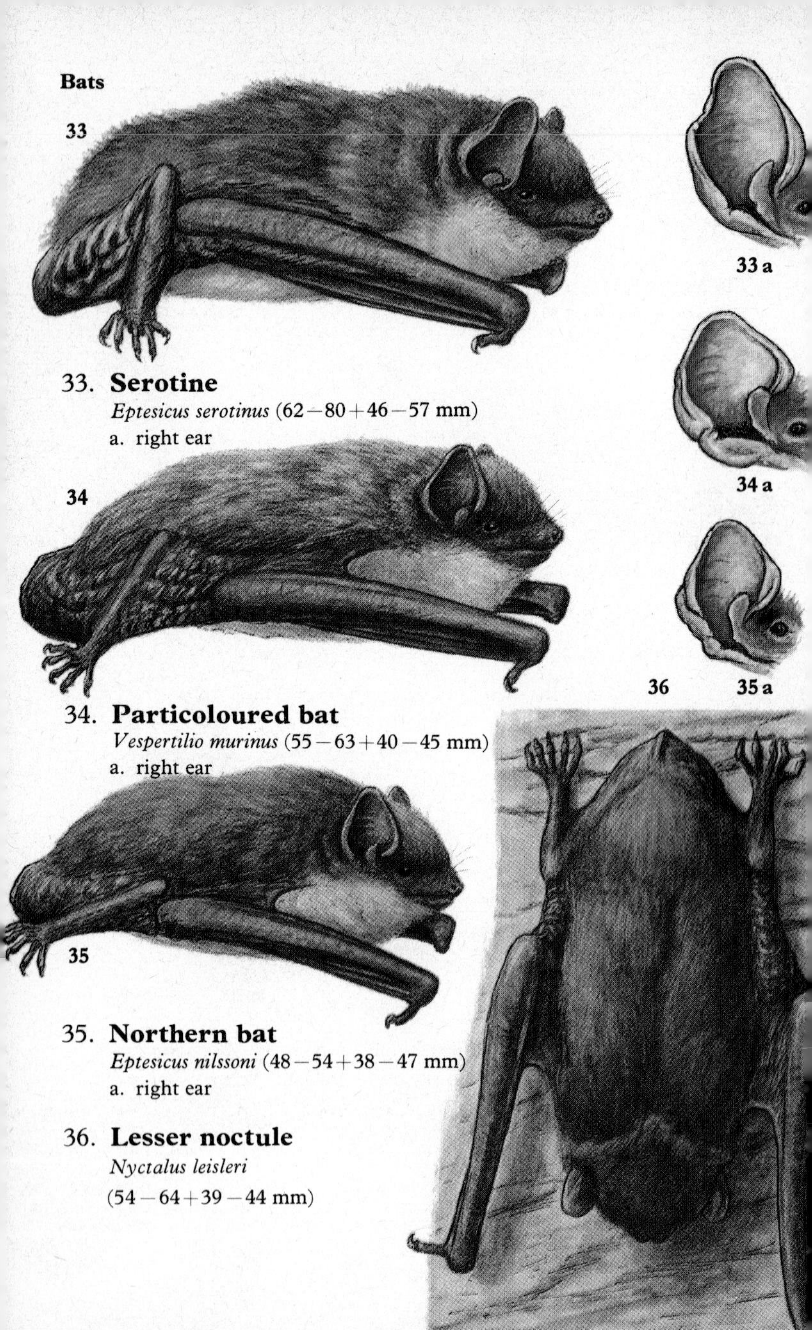

33. **Serotine**
Eptesicus serotinus (62–80+46–57 mm)
a. right ear

34. **Particoloured bat**
Vespertilio murinus (55–63+40–45 mm)
a. right ear

35. **Northern bat**
Eptesicus nilssoni (48–54+38–47 mm)
a. right ear

36. **Lesser noctule**
Nyctalus leisleri
(54–64+39–44 mm)

37 a

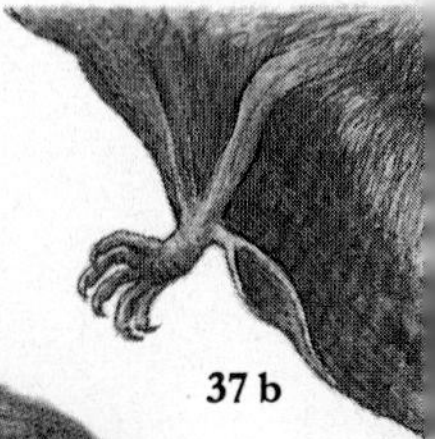

37 b

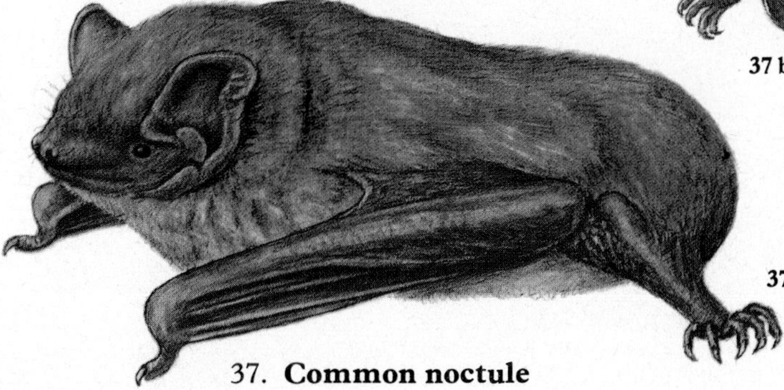

37

37. **Common noctule**

Nyctalus noctula (69 – 82 + 41 – 59 mm)

a. in flight

b. right hind-leg, from below

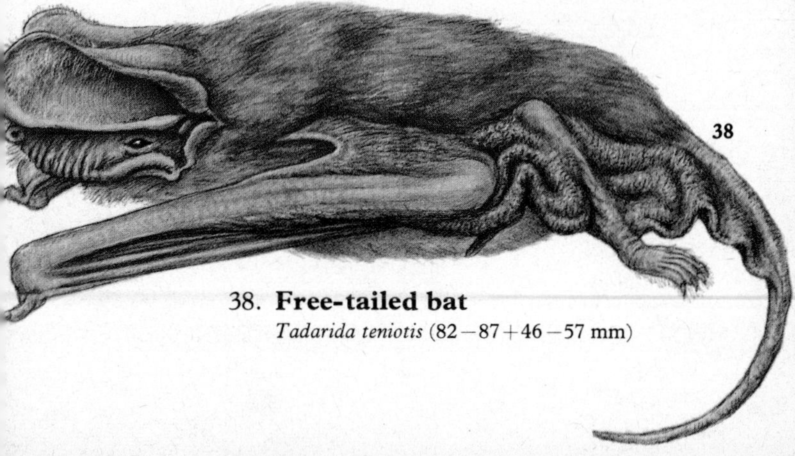

38

38. **Free-tailed bat**

Tadarida teniotis (82 – 87 + 46 – 57 mm)

39
39 a
40 a
40 b
40 d

39. Rabbit

Oryctolagus cuniculus
(34 – 45 + 4 – 8 cm)
a. new-born young

40. Brown hare

Lepus capensis (48 – 68 + 7 – 11 cm)
a. new-born leveret
b. skull
c. moving at speed
d. tracks made in c.
e. tracks when moving slowly

40

40 c

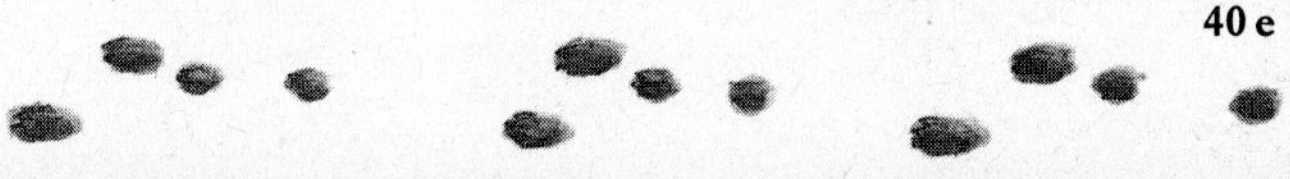

40 e

41. Mountain hare

Lepus timidus (46 – 61 + 4 – 8 cm)

a. winter coat

b. summer coat

41 b

42

42 a

42 b

42 c

42. Russian flying squirrel
Pteromys volans (15 – 17 + 9·5 – 13 cm)
a–c. successive stages in gliding

43. Grey squirrel
Sciurus carolinensis (24 – 30 + 20 – 25 cm)

44 a
44 f
44 g
44 h
44 b

44 d

44 c

44 e

44. **Red squirrel**

Sciurus vulgaris (19 – 28 + 14 – 24 cm)

a. British form in summer coat
b. black form
c. red-brown form
d. skull
e. tracks
f. spruce cone, g. pine cone and
h. hazelnut—all gnawed by Red squirrel

45. **European souslik**

Citellus citellus (19 – 22 + 5·5 – 7·5 cm)

46. **Spotted souslik**

Citellus suslicus (18·5 – 26 + 3·2 – 4 cm)

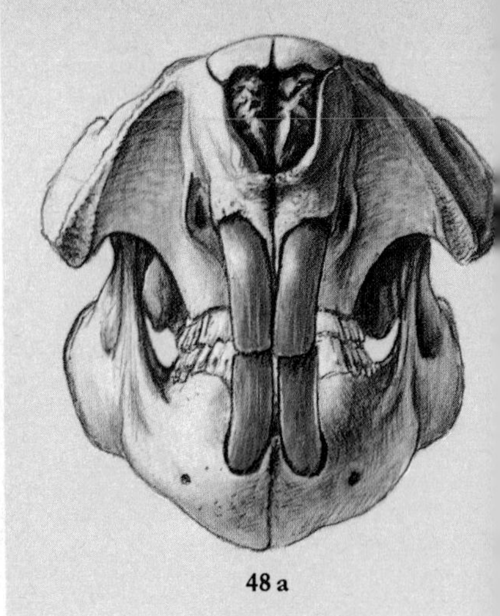

47

47. **Alpine marmot**
Marmota marmota (50·5 – 57·5 + 13 – 16 cm)

48. **European beaver**
Castor fiber (74 – 97 + 28 – 38 cm)
a. skull front view
b. swimming at the surface
c. swimming underwater

48 c
48 b
48

49. **Hazel dormouse**
Muscardinus avellanarius (6 – 9 + 5·7 – 7·5 cm)

50. **Forest dormouse**
Dryomis nitedula (8 – 13 + 8 – 9·5 cm)

51. **Garden dormouse**
Eliomys quercinus (10 – 17 + 9 – 12·5 cm)

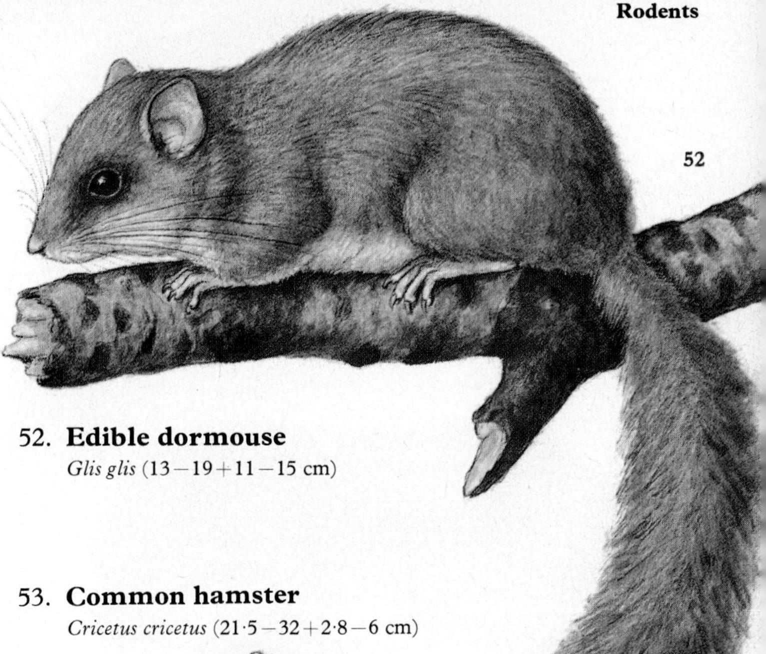

52

52. **Edible dormouse**
Glis glis (13 – 19 + 11 – 15 cm)

53. **Common hamster**
Cricetus cricetus (21·5 – 32 + 2·8 – 6 cm)

53

54. **Wood lemming**
Myopus schisticolor (8·5 – 9·5 + 1·5 – 1·9 cm)

55

55. **Norwegian lemming**
Lemmus lemmus (13 – 15 + 1·5 – 1·9 cm)
a. dark variety

55 a

56

56 b

56. Bank vole

Clethrionomys glareolus (8 — 12·3 + 3·6 — 7·2 cm)

a. cheek teeth, left upper jaw (above) and left lower jaw (below)
b. hazelnut opened by Bank vole

57. Northern red-backed vole

Clethrionomys rutilus (9·8 — 11 + 2·3 — 3·5 cm)

1
2
3
1
2
3
56 a

57

58. Grey-sided vole

Clethrionomys rufocanus (11 — 13 + 2·8 — 4 cm)

58

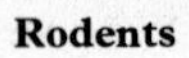

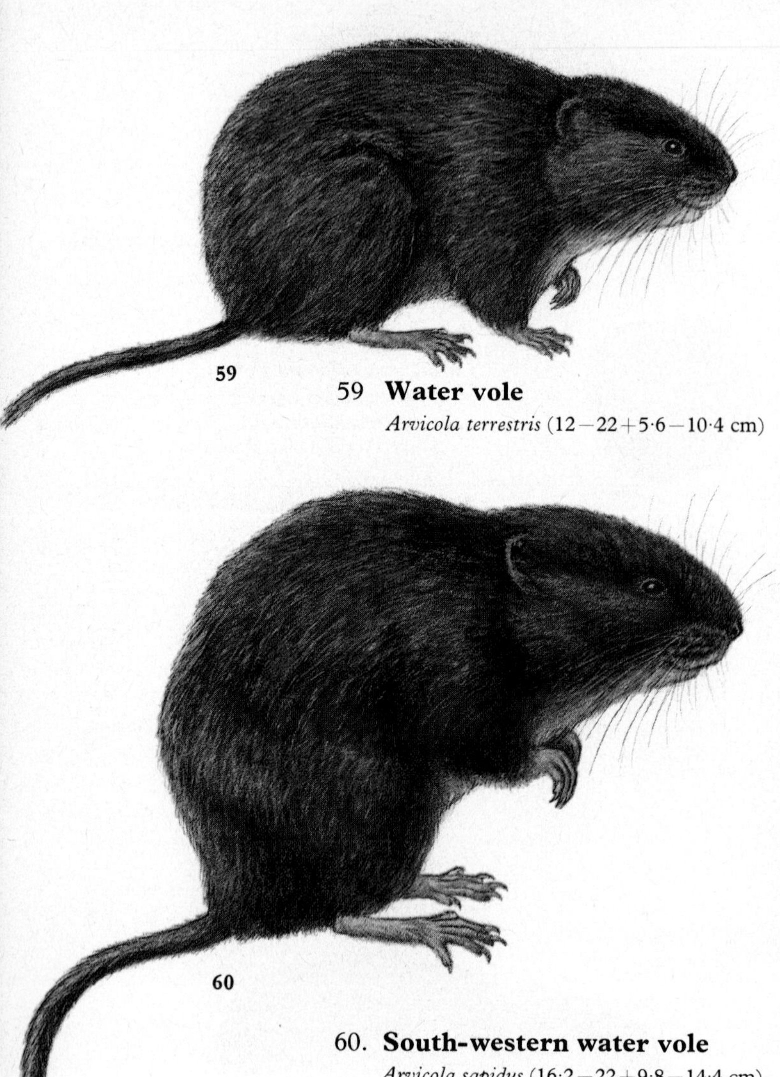

59 **Water vole**
Arvicola terrestris (12–22+5·6–10·4 cm)

60. **South-western water vole**
Arvicola sapidus (16·2–22+9·8–14·4 cm)

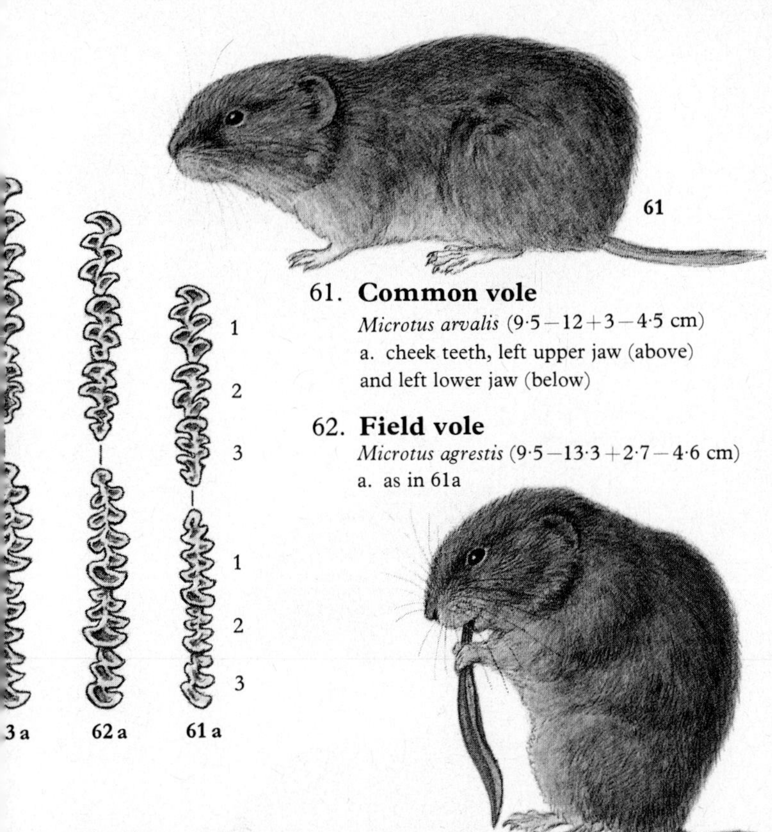

61. Common vole

Microtus arvalis (9·5 – 12 + 3 – 4·5 cm)
a. cheek teeth, left upper jaw (above) and left lower jaw (below)

62. Field vole

Microtus agrestis (9·5 – 13·3 + 2·7 – 4·6 cm)
a. as in 61a

63. Northern vole

Microtus oeconomus (11·8 – 14·8 + 4 – 6·4 cm)
a. as in 61a

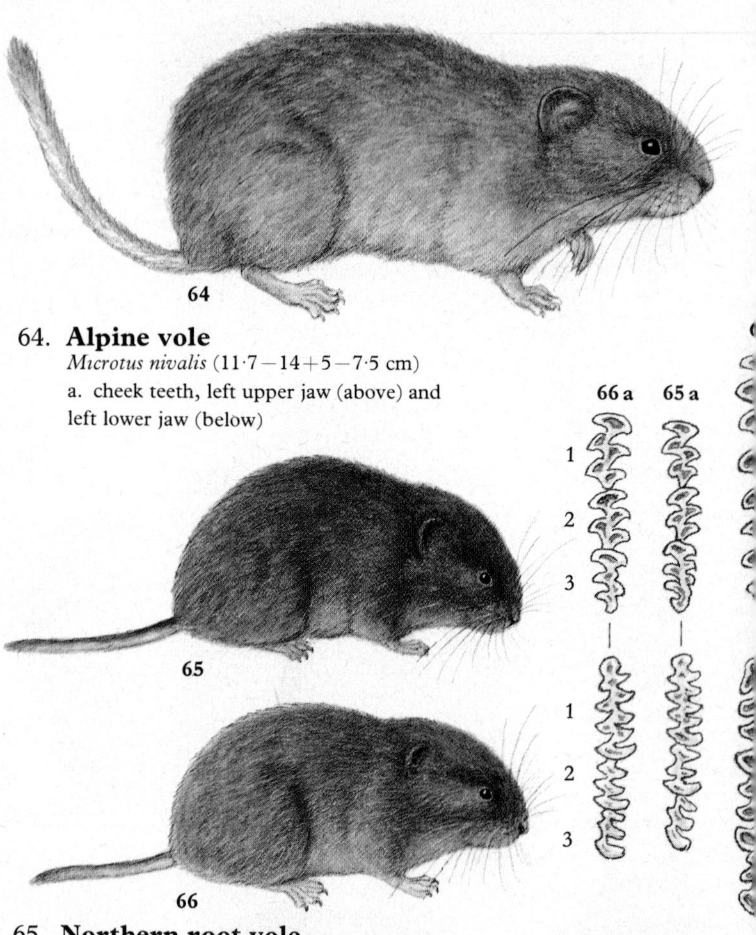

64. **Alpine vole**
Microtus nivalis (11·7 – 14 + 5 – 7·5 cm)
a. cheek teeth, left upper jaw (above) and left lower jaw (below)

65. **Northern root vole**
Pitymys subterraneus (7·5 – 10·6 + 2·5 – 3·9 cm)
a. as in 64a

66. **Mediterranean root vole**
Pitymys savii (8·2 – 10·5 + 2·1 – 3·4 cm)
a. as in 64a

67. **Iberian root vole**
Pitymys duodecimcostatus (9·3 – 10·7 + 2 – 2·9 cm)

67

68. **Musk-rat**
Ondatra zibethicus (26 – 40 + 19 – 27·5 cm)

69. **Mole-rat**
Spalax microphthalmus (18·5 – 27 + O cm), western form

70. **Harvest mouse**
Micromys minutus (5·8 – 7·6 + 5·1 – 7·2 cm)

71. **Striped mouse**
Apodemus agrarius (9·7 – 12·2 + 6·6 – 8·8 cm)

72. **Wood mouse**
Apodemus sylvaticus (7·7 – 11 + 6·9 – 11·5 cm)

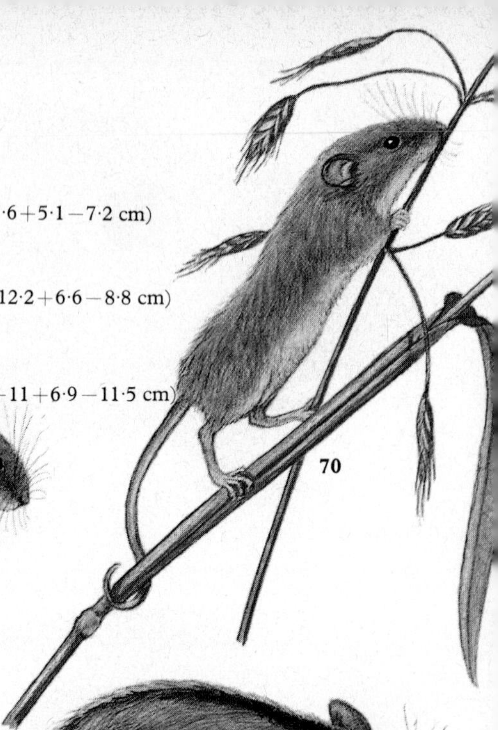

70

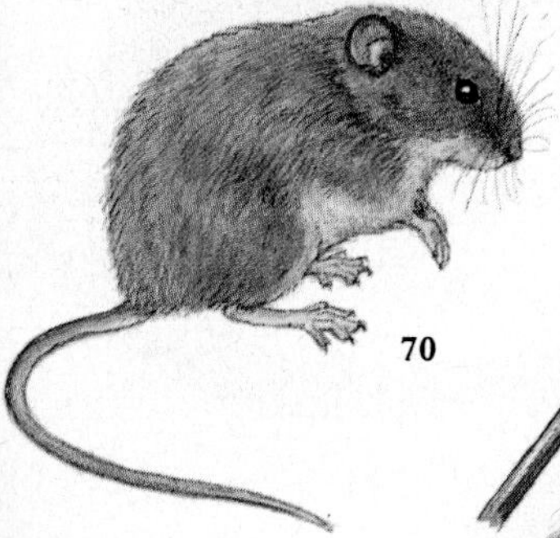

70

71

73. **Yellow-necked mouse**
Apodemus flavicollis 8·8 – 13 + 9·2 – 13·4 cm)
a. ventral view
b. upper incisor
c. spruce cone gnawed by Yellow-necked mouse
d. hazelnut opened by Yellow-necked mouse

72

73

73 c

74. House mouse

Mus musculus (7·2 – 10·3 + 5 – 10·2 cm), pale form

a. dark form

b. upper incisor

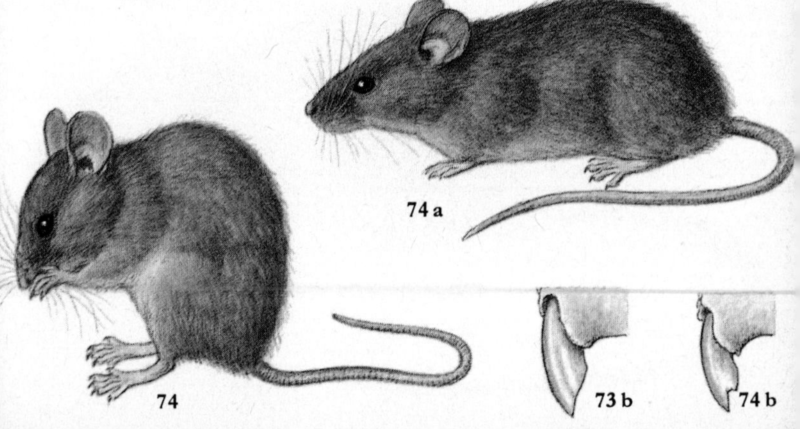

75 a

75. **Black rat**

Rattus rattus (15·8 – 23·5 + 18·6 – 25·2)
a. *alexandrinus* form
b. skull

75

76. **Brown rat**

Rattus norvegicus (21·4 – 27·3 + 17·2 – 22·9)
a. skull

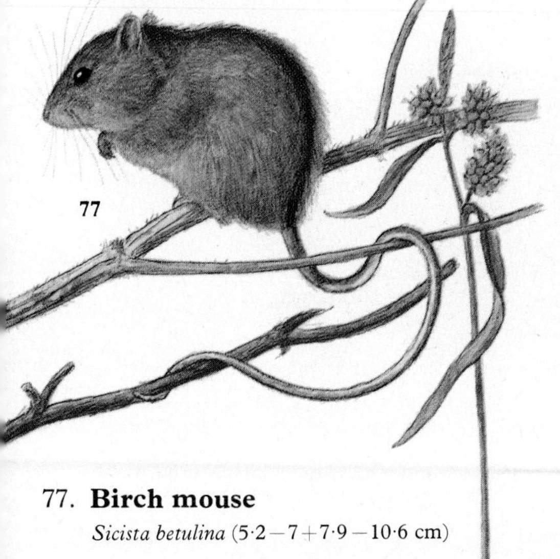

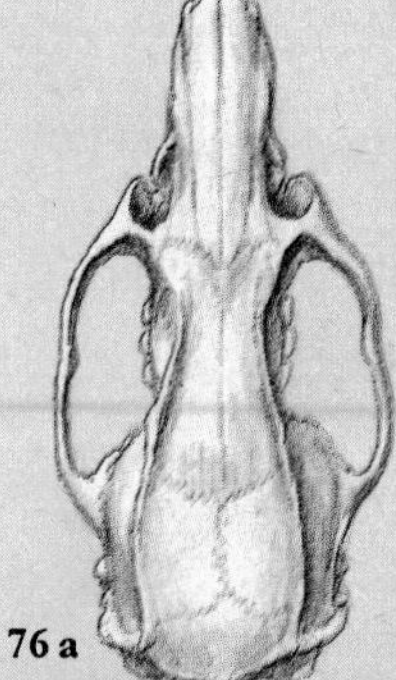

77. **Birch mouse**

Sicista betulina (5·2 – 7 + 7·9 – 10·6 cm)

78. **Crested porcupine**
Hystrix cristata (57 – 68 + 5 – 12 cm)

79. **Coypu**
Myocastor coypus (42 – 60 + 30 – 45 cm)

80. **European genet**
Genetta genetta (47 – 58 + 41 – 48 cm)

81. **Egyptian mongoose**
Herpestes ichneumon (51 – 55 + 33 – 45 cm)

84b

84c

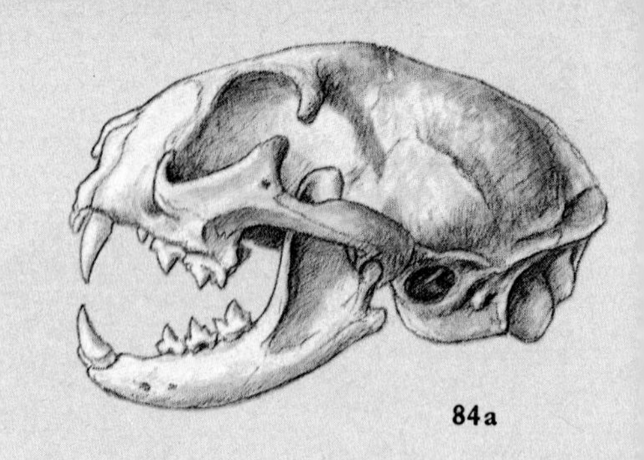
84a

82. **European lynx**
Felis lynx (80 – 130 + 11 – 25 cm)

83. **Spanish lynx**
Felis pardina (85 – 110 + 12 – 13 cm)

82

84. **European wild cat**
Felis silvestris (47 – 80 + 26 – 37 cm)
a. skull
b. footprint (fore-foot)
c. footprint (right hind-foot)

85

85. **Pine marten**

Martes martes (42 – 52 + 22 – 26·5 cm)

a. front view
b. upper jaw
c. footprint (left fore-foot)
d. two types of track

86. **Beech marten**

Martes foina (42 – 48 + 23 – 26 cm)

a. front view
b. upper jaw
c. footprint (left fore-foot)

85 d

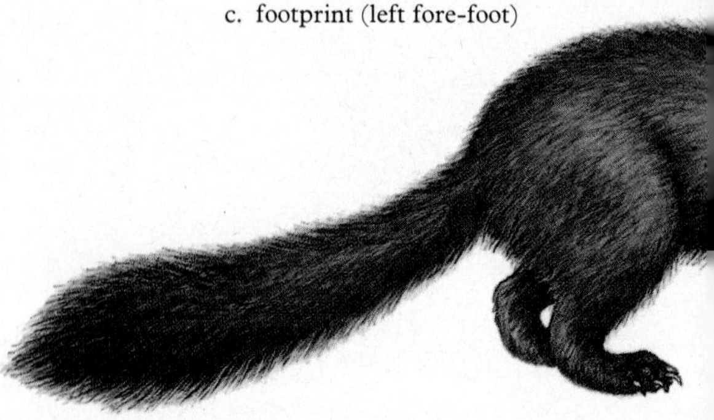

85 a

86 a

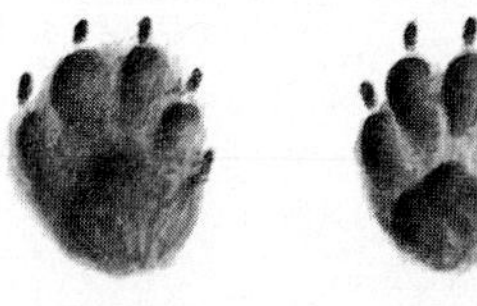

85 c

86 c

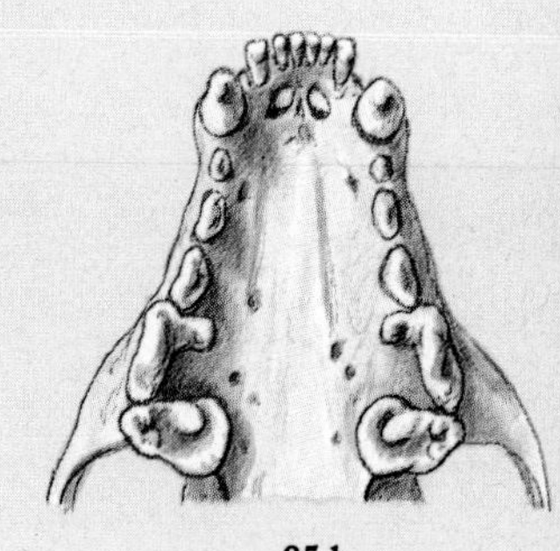

85 b

86

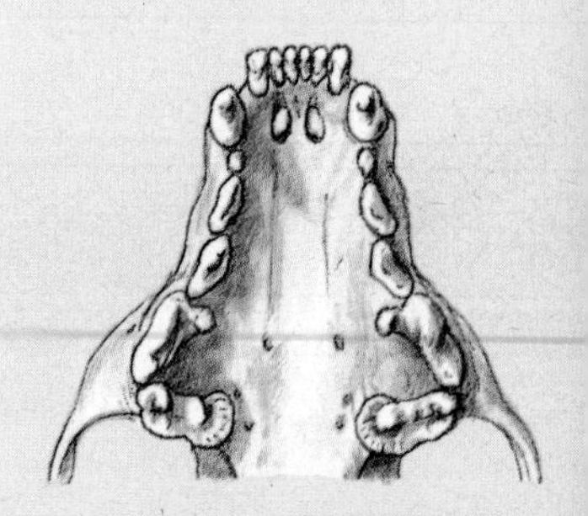

86 b

87
87 a
87 b

88

87. **Stoat**

Mustęla erminea (22 – 29 + 8 – 12 cm)

a. transitional coat
b. winter coat
c. footprint (left fore-foot)
d. typical bounding tracks (left);
other tracks (centre and right)

88. **Weasel**

Mustela nivalis (16 – 23 + 4 – 6·5 cm)

a. winter coat

89. **Least weasel**

Mustela rixosa (13 – 19·5 + 2·8 – 5·2 cm)

88 a

87 c

87 d

89

90

90. **European polecat**

Mustela putorius (32 – 45 + 13 – 19 cm)

a. ferret
b. footprint (left fore-foot)
c. different types of track

90 a

91. **European mink**

Mustela lutreola (28 – 40 + 12 – 15 cm)
a. head of European mink
b. head of North American Mink

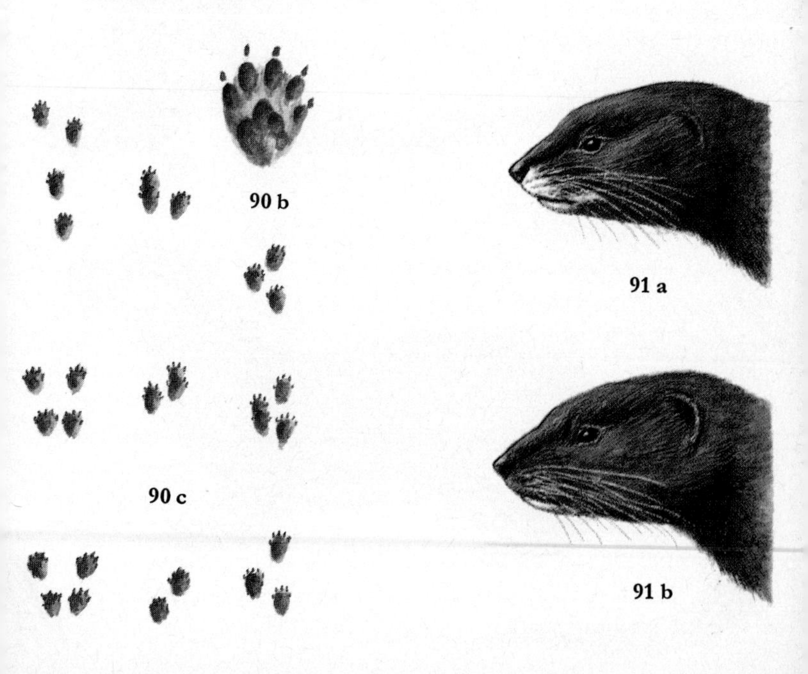

92. **Otter**
Lutra lutra (50 – 95 + 26 – 55 cm)
a. footprints; left fore-foot (left) and left hind-foot (right)

93. **Badger**
Meles meles (60 – 90 + 11 – 20cm)
a. skull
b. footprints: left fore-foot (above) and left hind-foot (below)

94. **Wolverine or Glutton**
Gulo gulo (70 – 83 + 16 – 25 cm)

92 a

92

92

93 a
93 b
93
94

95
97

95. Wolf

Canis lupus (110 – 140 + 35 – 50 cm)

a. footprint of left fore-foot

b. footprint of large domesticated dog

96. Asiatic jackal

Canis aureus (84 – 105 + 20 – 24 cm)

97. Raccoon-dog

Nyctereutes procyonoides (55 – 70 + 18 – 25 cm)

95 a

95 b

Carnivores
98 c
98 a
98 b
98
98

99

98. Common red fox

Vulpes vulpes (58–85+35–55 cm),
two colour variants

a. skull
b. footprint of left fore-foot
c. footprint of small domesticated dog

99. Arctic fox

Alopex lagopus (50–65+28–33 cm),
in winter

a. in summer
b. 'blue fox' in winter and summer

99 a

99 b

100. **Brown bear**

Ursus arctos (170 – 250 + 6 – 14 cm)

a. track

100 a

101. **Polar bear**
Thalarctos maritimus (160 – 250 + 8 – 10 cm)

102. **Raccoon**
Procyon lotor (48 – 70 + 20 – 26 cm)

103 c

103. Common seal

Phoca vitulina (145 – 195 + 7 – 9 cm)

a. swimming, seen from above
b. pup
c. characteristic resting positions
d. skull
e. right half of lower jaw
f. a cheek tooth

103 b

103 a

103

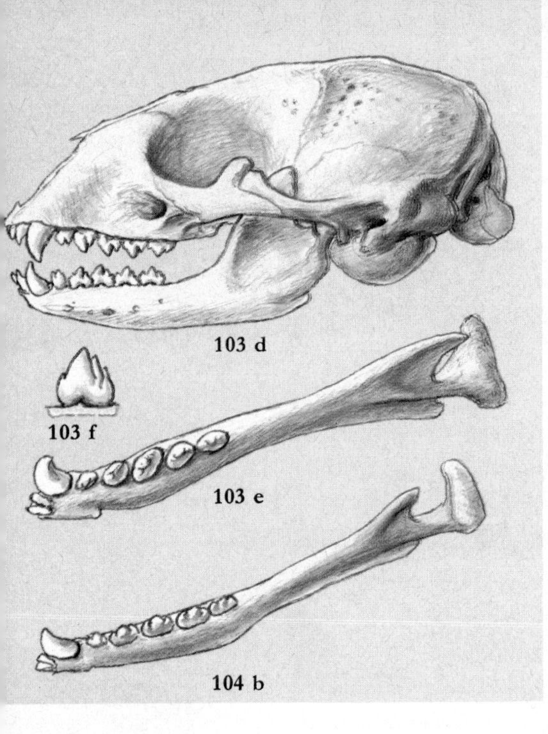

104. **Ringed seal**

Pusa hispida (100 – 185 + c.10 cm)
a. swimming, seen from above
b. right half of lower jaw

104 a

104

105. **Grey seal**

Halichoerus grypus (165 – 330 cm), male
a. female
b. pup
c. a cheek tooth

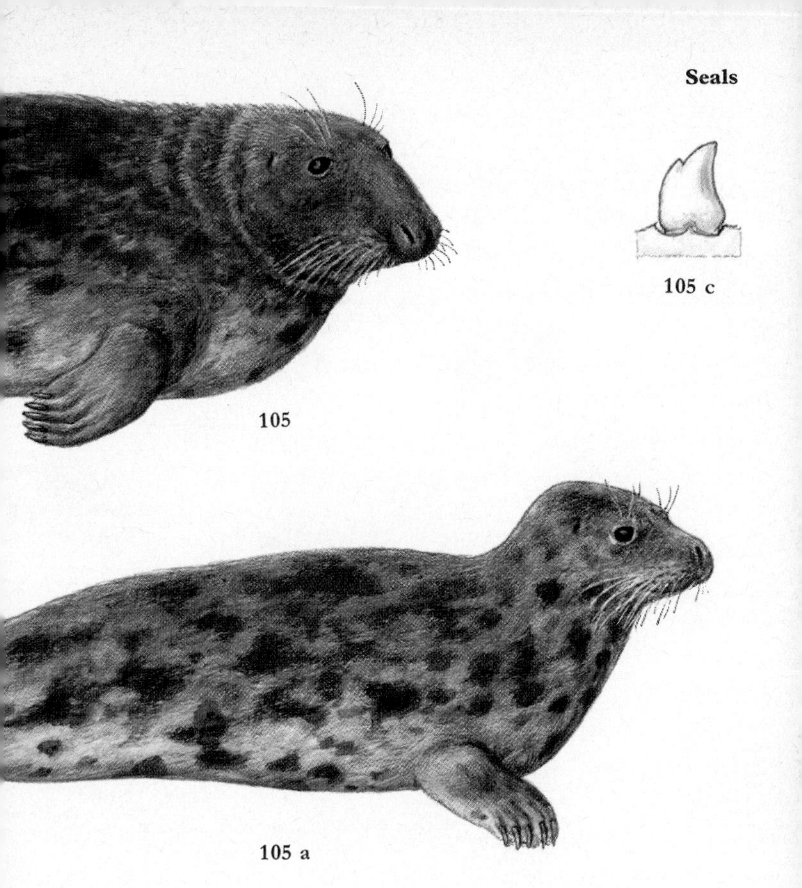
105 c

105

105 a

106. **Monk seal**

Monachus monachus (230 – 380 cm)

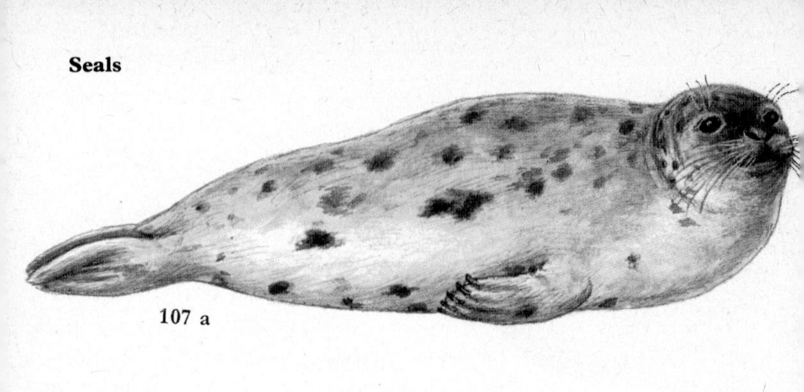

107 a

107 b

108. Hooded seal

Cystophora cristata (200 – 235 cm), male

a. with inflated nasal hood

b. pup

108

107. **Greenland or Harp seal**

Pagophilus groenlandicus (155 – 220 cm), male

a. 2 year-old juvenile

b. new-born pup

109

110. **Walrus**

Odobenus rosmarus (male 300 – 450 cm, female up to 300 cm)

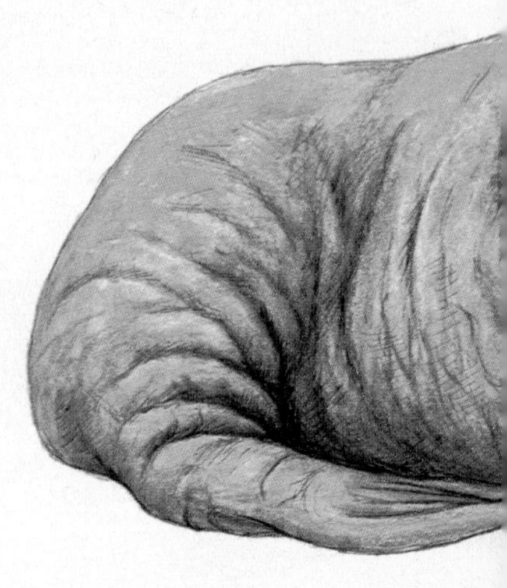

109. **Bearded seal**
Erignathus barbatus (220 – 310 cm)

111. Wild boar

Sus scrofa (110 – 155 + 15 – 20 cm),
female (sow)
a. male (boar)
b. young
c. footprint

111 a

112. Chinese water-deer

Hydropotes inermis (78–100+6–8 cm)

113. Chinese muntjac

Muntiacus reevesi (80–100+c.17 cm)

a. left antler

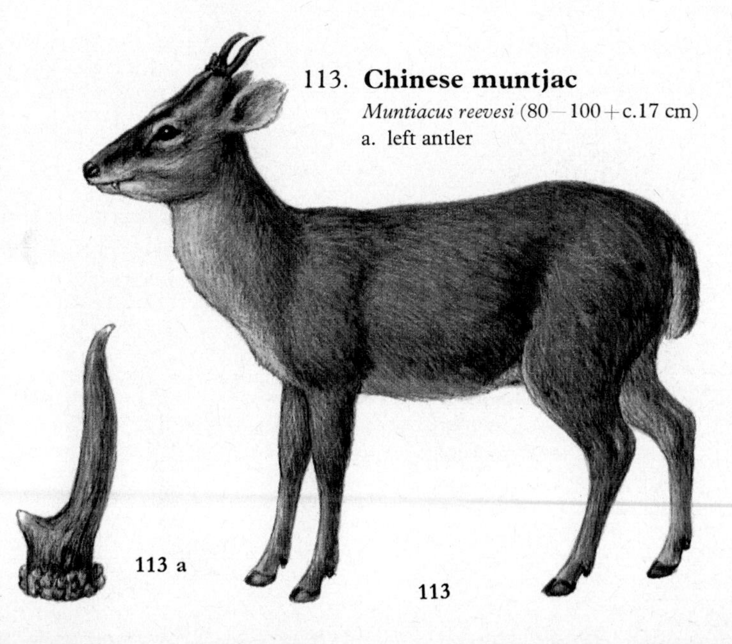

114 g

114 h

114. **Red deer**

Cervus elaphus (165 – 250 + 12 – 15 cm), male (stag)
a. with antlers in "velvet".
b-f. right antler with 1, 2, 3, 4 and 6 points
g. footprint in soft ground or when jumping
h. footprint on hard ground or when walking

114

114 a
114 b
114 c
114 d
114 e
114 f

115 a
115 b
115

115. **Red deer**

Cervus elaphus (165 – 250 + 12 – 15 cm), female (hind)
a. calf
b. rump patch

116 b

116. **Sika deer**

Cervus nippon (110 – 130 + 10 – 15 cm), male (stag) in summer
a. left antler
b. winter coat, rump patch

116 a

117. **Fallow deer**

Dama dama (130 – 160 + 16 – 19 cm), male) buck in winter

a. female (doe) in summer
b. female (doe) dark form
c. rump patch
d. footprint

117 d

117

117 c
117 b
117 a

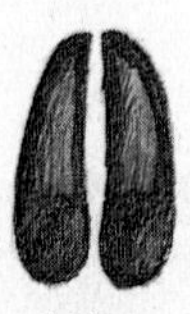

118 g

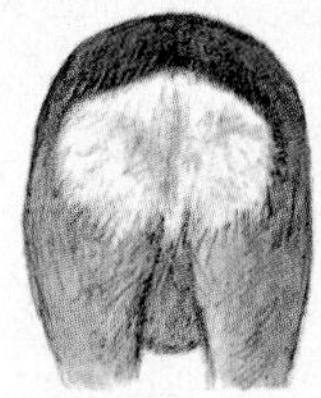

118 c

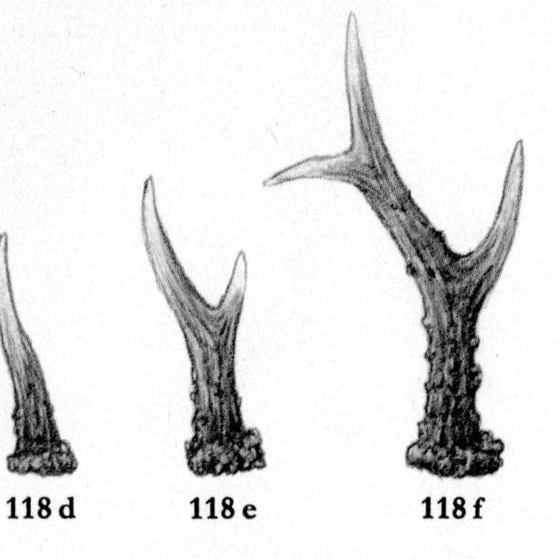

118 d 118 e 118 f

118

118. **Roe deer**

Capreolus capreolus (95 – 135 + 2 – 3 cm), male (buck) in winter

a. female (doe) in summer
b. kid
c. rump patch, on right when alarmed
d-f. right antler with 1, 2 and 3 points
g. footprint

118 b

118 a

119. **White-tailed deer**

Odocoileus virginianus (150 – 180 + 15 – 28 cm), male in winter
a. in summer, rump patch

119

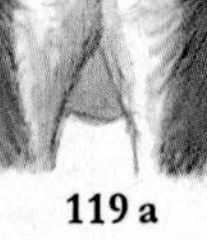

119 a

120 a

120 b

120. **Reindeer**
Rangifer tarandus (185 – 215 + c.15 cm), male (bull)
a. right antler in "velvet"
b. footprint

120

121. **Elk**

Alces alces (200 – 290 + 4 – 5 cm), male (bull)

a. branched antler (cervine type)
b. shovel-shaped antler (palmate type)
c. footprint

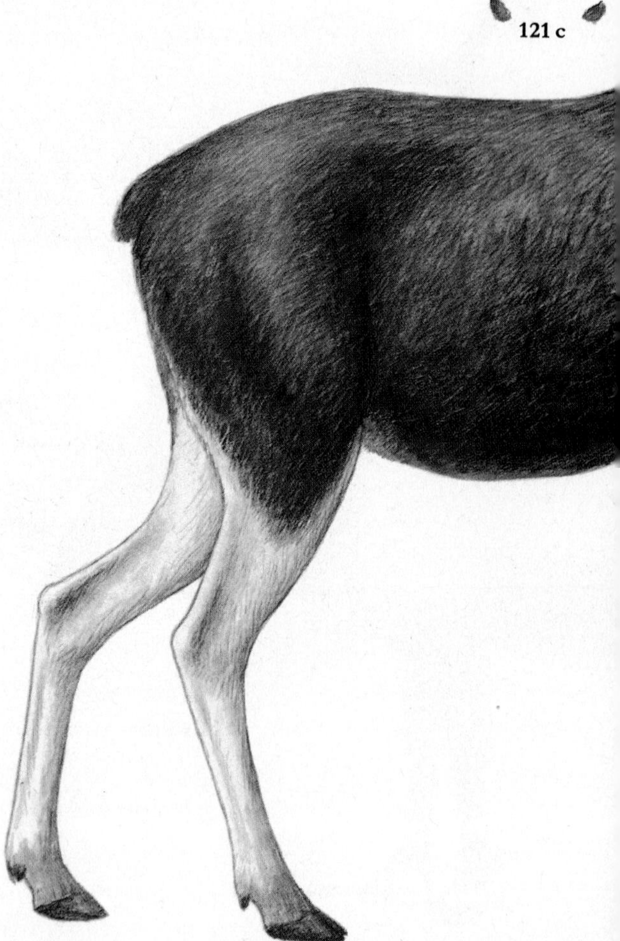

121 a
121 b

122. **European bison or Wisent**

Bison bonasus (270 + 80 cm), male (bull)

122

124 a
123

124

123. **Musk-ox**
Ovibos moschatus (200 – 245 + ca. 10 cm), (bull)

124. **Mouflon**
Ovis musimon (110 – 130 + 3 – 6 cm), male (ram)
a. female (ewe)

125
125 a
125 b

125. **Chamois**

Rupicapra rupicapra (110 – 130 + 3 – 4 cm), male in summer

a. female in winter
b. footprint

126. **Ibex**

Capra ibex (130 – 145 + 12 – 15 cm), male in summer

a. female in summer
b. footprint

126 b

126 a

126

127. **Barbary ape**
Macaca sylvana, (60 – 71 cm), male

128. **Porpoise**
Phocaena phocoena (1·35 – 1·85 m)
a. swimming, see p. 100
b. skull, see p. 100
c. head and flippers, dorsal view, see p. 100
d. tail flukes, dorsal view, see p. 100

129. **Common dolphin**
Delphinus delphis (1·80 – 2·60 m)

130. **Bottle-nosed dolphin**
Tursiops truncatus (2·80 – 4·10 m)

131. **White-sided dolphin**
Lagenorhynchus acutus (1·95 – 2·80 m)

132. **White-beaked dolphin**
Lagenorhynchus albirostris (2·35 – 3·10 m)

128
129
130
131
132

133. **Risso's dolphin**
Grampus griseus (2·5 – 4 m)

134. **Killer whale**
Orcinus orca (3·8 – 9·5 m), male

135. **Pilot whale**
Globicephala melaena (4·3 – 8·7 m)

136. **White whale**
Delpinapterus leucas (3·6 – 5·5 m)

137. **Narwhal**
Monodon monoceros (3·9 – 5·5 m), male

138. **Cuvier's beaked whale**
Ziphius cavirostris (5·5 – 9 m)

139. **Sowerby's whale**
Mesoplodon bidens (4·2 – 5·6 m)

137

138

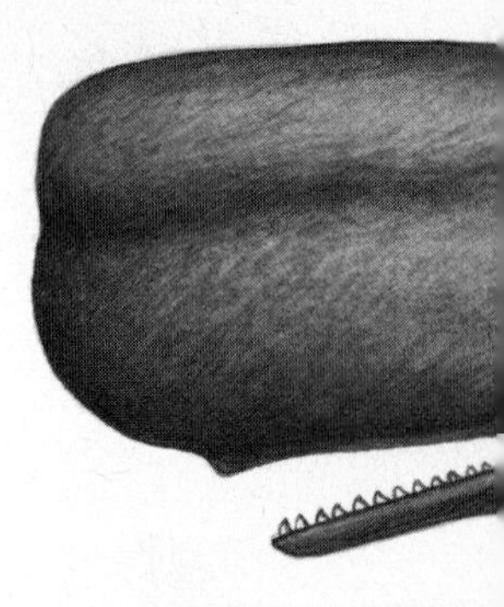

140. **Bottlenose whale**
Hyperoodon ampullatus (7 – 9·5 m)

141. **Sperm whale**
Physeter catodon (9 – 18 m), male
a. the 'blow', see p. 100

142. **Fin whale**
Balaenoptera physalus (18·5 – 25 m)
a. the 'blow', see p. 100
b. a baleen plate, see p. 100

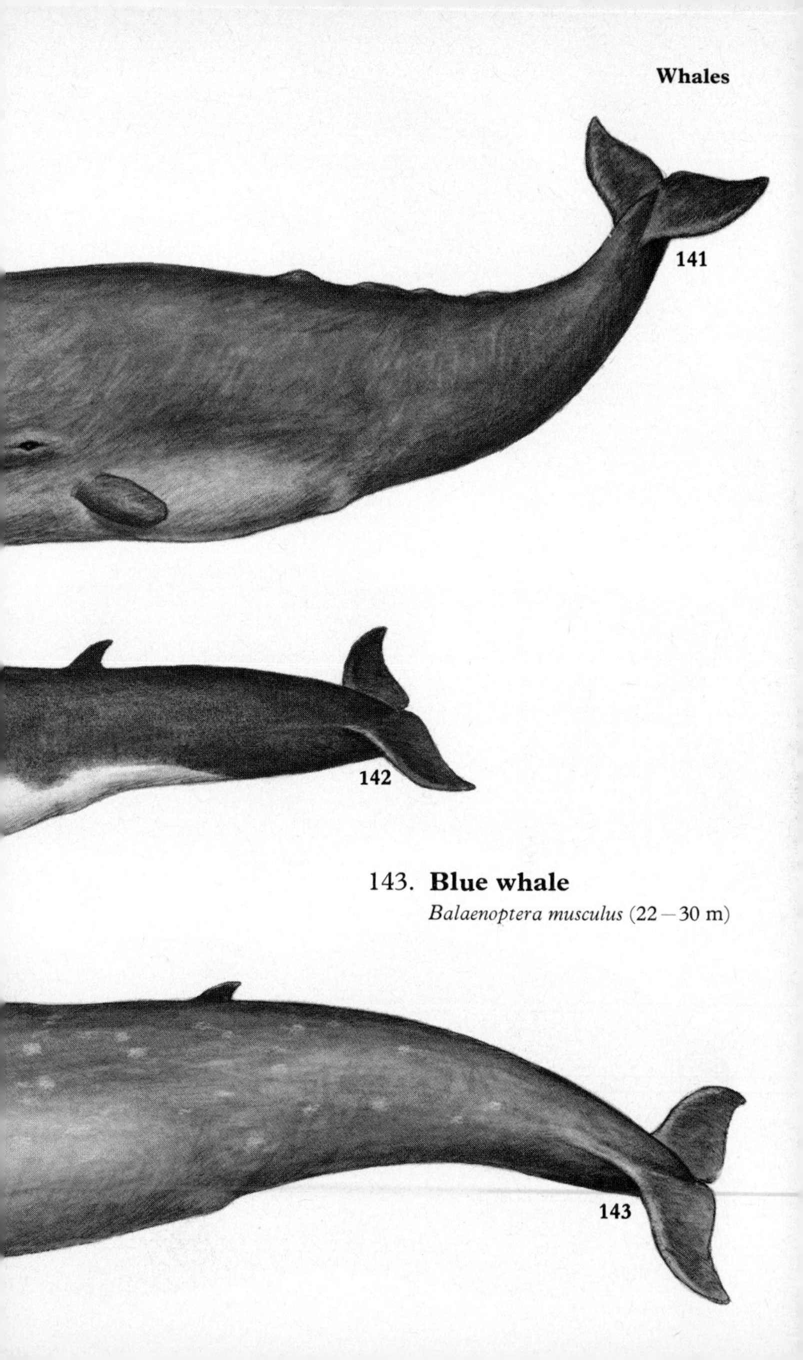

143. **Blue whale**
Balaenoptera musculus (22 – 30 m)

146

144. **Lesser rorqual**
Balaenoptera acutorostrata (8 – 10·5 m)

145. **Sei whale**
Balaenoptera borealis (12 – 18·5 m)

146. **Humpback whale**
Megaptera novaeangliae (11 – 16 mm)
a. the 'blow', see p. 100

147. **Black right whale**
Eubalaena glacialis (14 – 18 m)
a. the 'blow', see p. 100

148. **Greenland right whale**
Balaena mysticetus (15 – 21 mm)

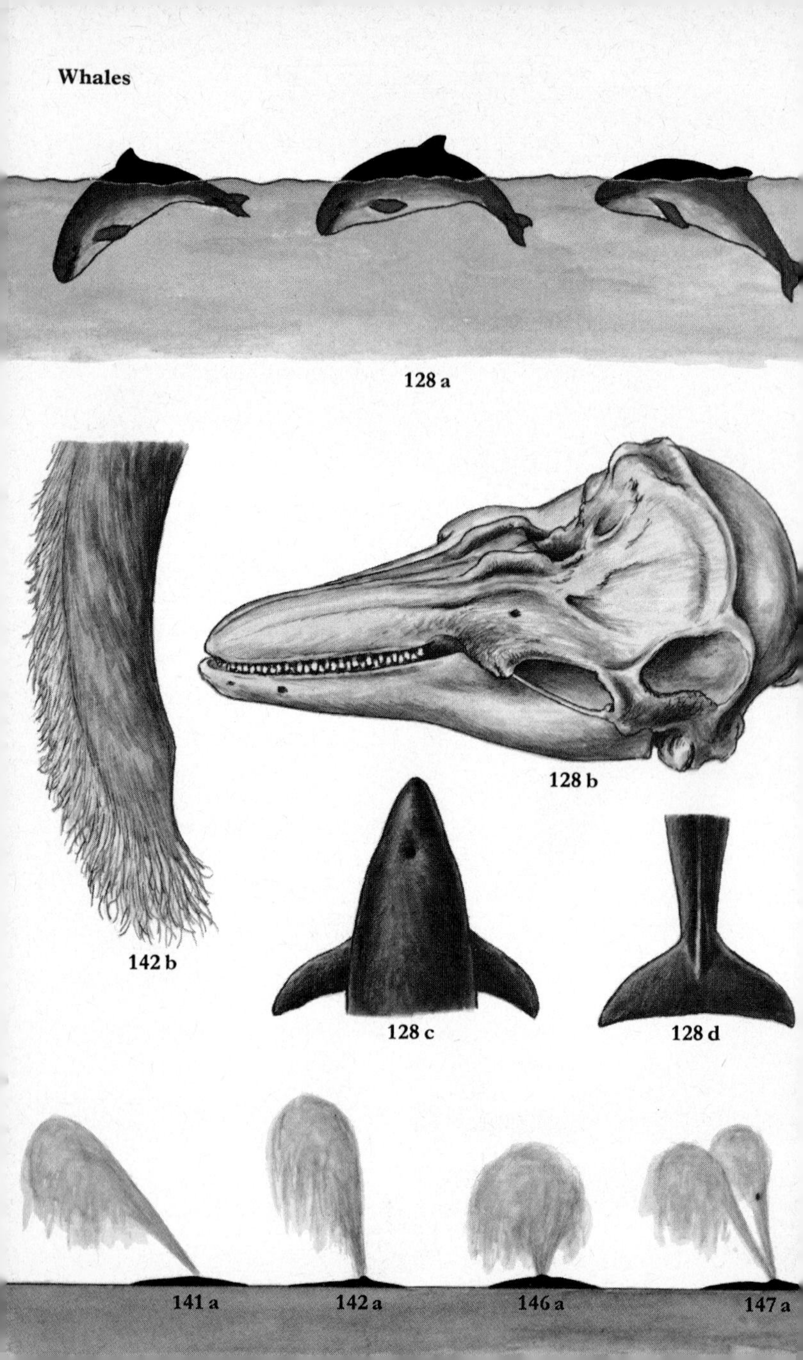
128 a
142 b
128 b
128 c
128 d
141 a
142 a
146 a
147 a

DESCRIPTIONS OF THE SPECIES

Insectivores

There are about 300 living species of insectivore (order Insectivora), distributed in all parts of the world except the Polar regions, Australia and South America; these are all that remain of a mammal group which at one time contained a large number of species. Structurally, they show several primitive characters: they have short legs, are plantigrade (i.e. they walk on the whole foot), have 5 clawed toes on both fore- and hind-feet, a tail that is often scaly and in certain species—including the mole—they have a dentition which is regarded as being the original complete mammal dentition. The latter consists of a total of 44 teeth: 3 incisors, 1 canine, 4 premolars and 3 molars in each half-jaw. The teeth form an unbroken series without any marked functional division, and the cheek teeth (premolars and molars) have several tips or cusps. Among other common characters of the insectivores, the muzzle is always elongated to some extent; the snout which is not unlike a proboscis may be movable and is always richly supplied with tactile organs. The ears and eyes are usually small, the latter particularly so and they may even be completely hidden in the fur. The coat is usually short and without any division into guard and woolly hairs; it may be silky-soft, as in the mole, or the hairs may be modified to form stiff, pointed spines, as in the hedgehog.

Most insectivores are terrestrial but a few are adapted to aquatic life. They are essentially predatory, feeding mainly on insects, and the smaller species require such a large amount of food that they must spend the greater part of the 24 hours actively searching for prey. Only one species hibernates: the hedgehog. The majority of insectivores have a relatively short or very short life but compensate for this by being prolific, often producing several litters of young in a year.

In Europe there are 16 species, most of which are small or very small, and these are classified into 3 clearly distinguished families: the hedgehogs (Erinaceidae), the shrews (Soricidae) and the moles (Talpidae).

Hedgehogs

1 Algerian hedgehog
Erinaceus algirus

Identification: Head is more slender than in the Northern hedgehog (2) and also more clearly outlined from the body; ears larger and broader. The spines on the upper surface of the head are divided into two, as though they had been 'parted' (1a); those on the back and the hairs on the belly are paler than in the Northern hedgehog. Body-length 20–25 cm (7–9 in.); tail about 3 cm ($1\frac{1}{4}$ in.); weight up to 850 g (1·9 lb).

Distribution: Belongs to a genus with species throughout Africa. In Europe, on the mainland, it is only found along the south and east coasts of Spain and France, but island populations are also present in the Balearics and the Canaries.

Habitat: Approximately the same as that of the Northern hedgehog.

Habits: Hibernation has not been recorded in the Algerian hedgehog.

Little is known about its habits, but they are probably much the same as those of the Northern hedgehog.

2 Northern hedgehog
Erinaceus europaeus

Identification: Head not sharply differentiated from the stout body, which has about 16,000 spines on the back. These are 2 cm ($\frac{3}{4}$ in.) long, dark brown or black, with white at each end. The head and belly have long, stiff hairs which are yellowish-white to dirty brown. Some specimens are very pale and albinos have been recorded. The pointed muzzle is always moist. Eyes black and comparatively large. Fore-feet larger than the hind-feet. Body-length 23–30 cm (9–12 in.); tail about 2·5 cm (1 in.); weight 450–1200 g (1–2$\frac{1}{4}$ lb). The Northern hedgehog has several forms. These differ from the Algerian hedgehog (1a) in the pattern of the 'hair-line' above the forehead. The specimen from Denmark (2a) is whitish on the belly (2b). Specimens from the British Isles are usually more uniform brown on the belly.

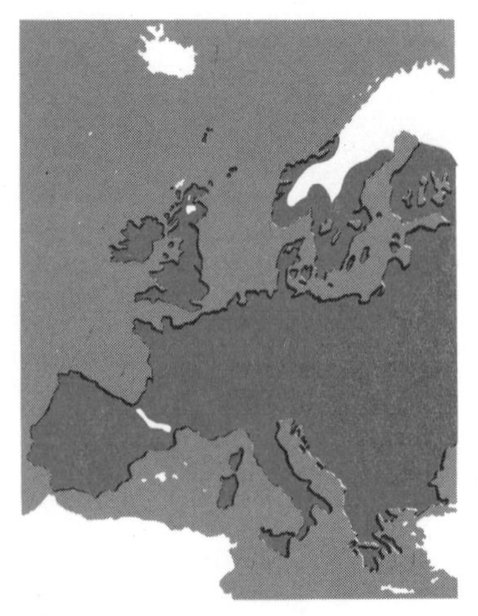

Distribution: Found throughout Europe—eastwards through northern Asia to the Pacific—except in northern Norway, Sweden and Finland, but is spreading steadily northwards, particularly in coastal regions.

Habitat: Common in most areas where there is some cover, including open country and suburban areas where there are gardens and parks; also in hedgerows, scrub and open woodland.

Habits: Mainly crepuscular and nocturnal. There are three main periods of activity in the 24 hours period: between 1800 and 2000 hours, between 0001 and 0200 hours and again early in the morning between 0400 and 0600 hours. It is almost omnivorous but insects, worms and snails are the main items of diet; berries and fruits, frogs, toads, reptiles, birds' eggs and young are also eaten. The daily requirement is about 200 g ($\frac{1}{2}$ lb). Hedgehogs find food by rooting about among fallen leaves, snuffling and snorting loudly as they search. Most of the daylight hours are spent in a well-concealed nest of dry leaves, grass and moss.

Mating takes place in April–May, shortly after the hedgehogs have come out of hibernation. It is not uncommon, early in the evening, to see and hear the male snuffling in circles around the female who adopts a defensive position, presenting her front to the male until mating actually takes place. The female (or sow) has a gestation period of 4–5 weeks and produces

–7 young. The new-born young are bout 5 cm (2 in.) long, blind and aked on the belly. The back has short vhite spines which are quickly eplaced by the typical brown spines. After 3–4 weeks the young begin to eave the nest. They are fully grown at months and become sexually mature he following summer. A second litter nay be produced later in the summer. n autumn, the hedgehog prepares a vell-lined nest on the ground under over, such as a pile of brushwood or n a compost heap. When the cold veather comes the animal curls up in his winter home and starts to hiber-ate. When the external temperature is –10° C (46–50° F) the hedgehog's ody temperature falls to about 6° C 43° F), a process which is under hor-nonal control. The hedgehog's worst nemy is man; a great number are illed on the roads. In Denmark, for nstance, it is estimated that about 20,000 hedgehogs are killed by cars very year. Large birds of prey, owls nd some carnivores such as the adgers, foxes and martens, attack and ucceed in killing hedgehogs in spite of he spines. The hedgehog's ability to url up when aware of danger is due to layer of muscles along the back and ides of the body. When these muscles ontract the rest of the body is com-ressed, curled into a ball, and the pines are erected. If the muscles are ontracted suddenly the spines can nflict unpleasant wounds on some redators, particularly on a dog's ose.

hrews

Members of the shrew family, the oricidae, are small, short-legged, nouse-like animals with a long, pointed muzzle. The eyes are small and vision plays only a minor role in orientation. The tail is usually longer than half the body-length. There is a scent gland on each side of the thorax or abdomen. Shrews are active throughout the year. The European species are classified in 4 genera: *Sorex*, *Neomys*, *Suncus* and *Crocidura*. In the first 2 genera the teeth have red-brown enamel at the tips, whereas in the other 2 the teeth are completely white.

3 Common shrew
Sorex araneus

Identification: The brownish-black coloration of the back extends far down the sides and is separated from the paler belly by a very broad transition zone. Body-length 70–87 mm ($2\frac{3}{4}$–$8\frac{1}{4}$ in.), tail relatively short, 35–45 mm ($1\frac{3}{8}$–$1\frac{3}{4}$ in.), and thin, with short hairs; hind-foot 11–13 mm (about $\frac{1}{2}$ in.) long; weight 3·5–14·0 g

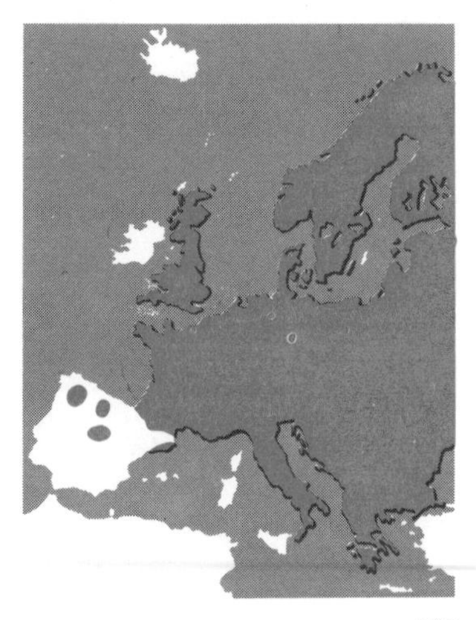

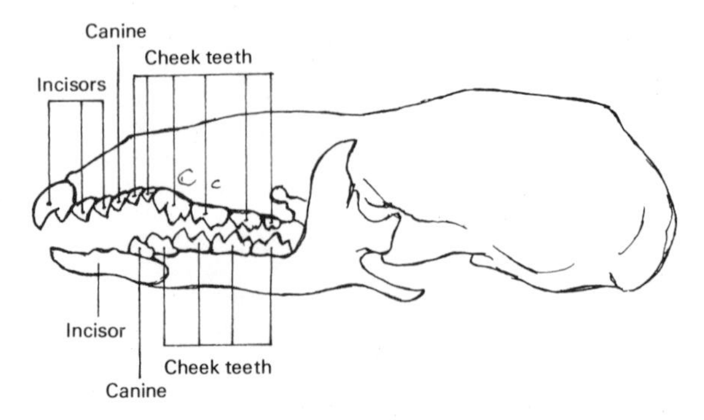

Skull of one of the shrews in the genus *Sorex* which has a total of 32 teeth. Th front incisor in the upper jaw is very large and has two points. This is followed b 5 small teeth: the second and third incisors, the canine and the first 2 cheek teeth Behind these are the remaining 4 cheek teeth which are larger. The lower jaw has single, large projecting incisor, a canine and finally 4 cheek teeth. In the gener *Neomys* (Water shrew) and *Suncus* the last intermediate tooth in the upper jaw i lacking, thus in these 2 genera the skull has a total of 30 teeth (see fig. 8c). Ther has been a further reduction in the genus *Crocidura* in which the first 2 small chee teeth are lacking (see figs. 10a and 12b).

($1\frac{1}{4}$–$4\frac{3}{4}$ oz), usually 6–10 g (2–$3\frac{1}{2}$ oz). Teeth: see fig. 3b. Note that the second and third incisors in each half of the upper jaw are about the same size and considerably larger than the canine, and that the 2 front cheek teeth are small. The lower canines are pointed.

Distribution: Throughout Europe, except Ireland, the greater part of the Iberian Peninsula, Corsica, Sardinia, Sicily and other Mediterranean islands; in Scandinavia it extends up to the Arctic Ocean and eastwards through northern Asia to Lake Baikal.

Habitat: Undemanding in choice of habitat. It occurs practically everywhere with low ground cover—in swamps, marshes and scrub as well as meadows and woods. Lives mainly among dense, low vegetation.

Habits: Active day and night and at al times of the year. It has about 1 periods of activity during the 24 hours, separated by short periods o rest; the peak of activity occurs a night. The daily intake of food is enormous: between half and three-quarters of the animal's own body-weight females suckling young require twice as much. Food is found with the hel of the long tactile hairs on the muzzle which is constantly moved from side to side as the shrew moves forwards by fits and starts. Vision is poor, hearing and tactile senses are good and the sense of smell help to determine whether or not the prey is edible. Large prey, such as a worm or a big beetle is handled in much the same way as a terrier treats a rat: it is shaken and torn into pieces with the pointed teeth, while the fore-feet hold the prey down. Water is often drunk and as it is swallowed the head is lifted, just as a bird raises its bill when swallowing. Because the food requirement is so great, shrews often die in traps that have not been visited for a few hours.

Long burrows are dug in loose earth immediately below the surface. When digging, the muzzle is used like a boring tool while the front paws move the loosened soil backwards. On the ground long surface runs are made in the grass or among leaves. The route system is complex and runs often cross each other as the animal searches for food; these may also be crossed by the runs made by mice or moles on similar forays. Shrews swim and climb well. Diet consists of insects, larvae, earthworms, spiders, woodlice and other small invertebrates. Fresh carrion and seeds containing oil, the latter especially in winter, are also eaten.

Two types of sound are made: a subdued twittering and a high-pitched screaming. The latter is aggressive and is heard when two individuals meet. Except during the periods of mating the shrew is strictly solitary and maintains a territory. The animals become sexually active after the winter, often by mid-April, and this continues until the autumn. The nest is built either below ground or at the surface hidden in low cover. The gestation period is 13–19 days and the female usually becomes pregnant again immediately after the birth of the first litter. A single female may produce 3 or more litters a year. The number of young per litter varies during the course of the season. In May there may be about 8 but this falls in August to about 5. At birth the young are about the size of a pea and weigh approximately 0·5 g ($\frac{1}{60}$ oz). They are suckled for about 3 weeks, by which time they have reached a weight of about 7 g ($\frac{1}{4}$ oz). The parent shrews usually die during late summer or early autumn. The life-span therefore does not exceed $1\frac{1}{2}$ years. Some of the young females may breed during their first summer. It is only young shrews which survive the winter. These are active throughout the winter, moving about in the soil or below the snow, but they may also enter barns and similar outbuildings.

Shrews are caught by many carnivores, also by owls and birds of prey. On account of their musky smell which comes from lateral gland, many predators seem to find them unpalatable and it is only owls, particularly Barn owls, which appear to eat them to any great extent.

4 Pygmy shrew
Sorex minutus

Identification: Back uniform brownish-grey. This colour extends far down the sides and there is no transitional zone (or only a faint one) between the back and the paler belly. Body-length 43–60 mm ($1\frac{3}{4}$–$2\frac{1}{4}$ in.) long; tail relatively long and thick, 31–46 mm ($1\frac{1}{4}$–$1\frac{3}{4}$ in.), and clothed with longer and denser hairs than in the preceding species. Hind-foot 9·5–11·0 mm ($\frac{3}{8}$–$\frac{7}{16}$ in.); weight 2·5–6·0 g ($\frac{1}{8}$–$\frac{1}{5}$ oz). The teeth are shown in fig. 4a. Note that the canines in the upper jaw are almost exactly the same size as the second and third incisors, whereas the first 2

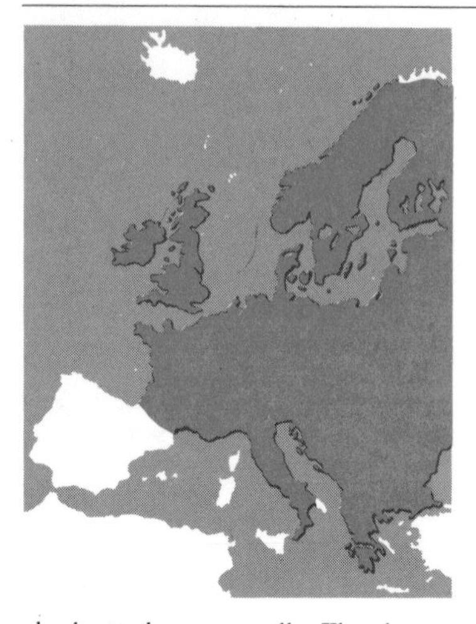

cheek teeth are small. The lower canines are pointed.

Distribution: Found throughout Europe, except the Iberian Peninsula, Corsica, Sardinia, Sicily and other Mediterranean islands. This species occurs throughout Scandinavia, except along the coasts of the Arctic Ocean, and extends eastwards to the Pacific.

Habitat: Often found in the same place as the Common shrew but usually present in smaller numbers.

Habits: In general, the same as those of the preceding species but with several interesting differences. The Pygmy shrew has significantly more periods of activity, with alternating rest periods, during the 24 hours than the Common shrew. Also, there are probably certain differences in relation to choice of food. The Pygmy shrew is apparently unable to dig its own runs in the soil but uses the tunnel systems and runways of other animals. Finally, its movements are considerably faster than those of the Common species, thus enabling it to avoid the latter. Trap investigations in England have shown that the ratio of the numbers of Pygmy shrew and Common shrew, caught in different localities, varies between 1:25 and 1:6. The lowest population density is found in woodland, the highest in heathland and similar habitats with tall, dense grass vegetation.

5 Laxmann's shrew
Sorex caecutiens

Identification: The brownish-black colour of the back does not extend far down the sides, thus the pale colour of the belly is more conspicuous than in the two preceding species. Body-length 44–67 mm ($1\frac{3}{4}$–$2\frac{5}{8}$ in.); tail relatively short and thick, 31–44 mm long ($1\frac{1}{4}$–$1\frac{3}{4}$ in.), the white underside sharply demarcated from the brown

upperside, and the tip has a tuft of hairs; hind-foot is 10·5–12·0 mm ($\frac{7}{16}$–$\frac{1}{2}$ in.) long; weight 3·5–7·0 g ($\frac{1}{6}$–$\frac{1}{3}$ oz). Teeth: see fig. 5a.

Distribution: Found in the northern part of the Soviet Union and westwards into North Sweden and northern Norway; in Finland, it is absent from the westernmost and southernmost districts. It also occurs in Poland.

Habitat: Found in woodland, especially in areas of coniferous forest, often in the vicinity of mountains. Occurs together with the Common and the Pygmy shrew. In the Soviet Union also found in the tundra, where it lives on high ground with plenty of cover.

Habits: Not significantly different from those of the Common shrew.

6 Alpine shrew
Sorex alpinus

Identification: The largest of the European shrews. Almost uniformly grey-black, the belly only slightly paler. Body-length 62–77 mm ($2\frac{1}{2}$–3 in.); tail relatively long, 62–75 mm ($2\frac{1}{2}$–$2\frac{3}{4}$ in.), two-coloured (pale underside) with black hairs; hind-foot 16–20 mm ($\frac{5}{8}$–$\frac{3}{4}$ in.) long; weight 9–23 g ($\frac{1}{3}$–$\frac{3}{4}$ oz). Teeth: see fig. 6a for comparison with other species of *Sorex*. Note that the 2 foremost cheek teeth in the upper jaw are relatively better developed than in the other species and that the lower canines have two points.

Distribution: Found in various mountain regions in central and eastern Europe and also in the Pyrenees.

Habitat: Not common in its distribution area, where it may be found together with the Common shrew. Extends up to altitudes of 3,000 metres (9,800 ft) and lives on the outskirts of coniferous forests, in scrub above the tree-line, and along streams; has also been recorded at much lower altitudes in coniferous forests.

Habits: Probably not significantly different from those of the other species of *Sorex*.

7 Least shrew
Sorex minutissimus

Identification: Body-length 39–53 mm ($1\frac{1}{2}$–$2\frac{1}{8}$ in.); tail 23–29 mm ($\frac{7}{8}$–$1\frac{1}{8}$ in.), densely haired; hind-foot 7·5–8·8 mm (*c.* $\frac{5}{16}$ in.); weight not over 4 g ($\frac{1}{5}$ oz), usually about 2 g ($\frac{1}{10}$ oz). The fur on the back is up to 3 mm ($\frac{1}{8}$ in.) long (at least twice as long in the other European shrews).

Distribution: Widespread in the northern parts of the Soviet Union, eastwards through northern Asia to the Pacific Ocean. This species also extends westwards into Finland and has been recorded on a few occasions in Northern Sweden and in Norway near Trondheim.

Habitat: Mainly in damp coniferous forests and in the vicinity of marshland.

Habits: Little information available. Observations on a small number of young individuals in Finland have shown that the activity of this species is about twice as great as that of the Common shrew. The observed animals were active for a total of 4 hours in a 24-hour period, as against 2 hours in the larger species, and the periods of activity were spread over the whole 24 hours. The food requirements of these animals—each individual weighed about 2 g ($\frac{1}{10}$ oz)—amounted to about 6 g ($\frac{3}{10}$ oz) in the 24 hours. They avidly ate mouse flesh as well as insects, including wood ants, and other arthropods. Weighing so little this shrew is a better climber than the other species and it also swims well.

Sorex isodon

Identification: Very similar to the Common shrew (3), but a little larger. Body-length 57–82 mm ($2\frac{1}{4}$–$3\frac{1}{4}$ in.); tail 41–55 mm ($1\frac{5}{8}$–$2\frac{1}{4}$ in.); hind-foot 13–15 mm (about $\frac{1}{2}$ in.). Back dark brown to blackish-brown without any sharp contrast with the belly which is grey-brown to yellow-brown. In the upper jaw the second cheek tooth is larger than in the Common shrew and like the other teeth has brown cusps.

Distribution: From the coasts of the Pacific Ocean westwards to eastern Finland, where there are isolated populations.

Habitat and habits: Lives mainly in mixed forest with some spruce; also found in houses. Its habits are the same as those of the Common shrew.

8 Water shrew
Neomys fodiens

Identification: Back very dark (slate-grey to almost black), sometimes with a brownish tinge; belly usually whitish and sharply demarcated from the dark back, but it may also be dark. There are often areas of white hairs near the eyes and ears. The tail is uniform in colour, the two fringes of long hairs on the underside being silver-grey. Body-length 72–96 mm ($2\frac{7}{8}$–$3\frac{3}{4}$ in.); tail 47–77 mm ($1\frac{7}{8}$–3 in.); hind-foot 16–20 mm ($\frac{5}{8}$–$\frac{3}{4}$ in.); weight 10–23 g ($\frac{1}{3}$–$\frac{2}{3}$ oz). Both the fore- and hind-feet have a fringe of hairs (8a). Muzzle—broader than in the species of *Sorex*, with the

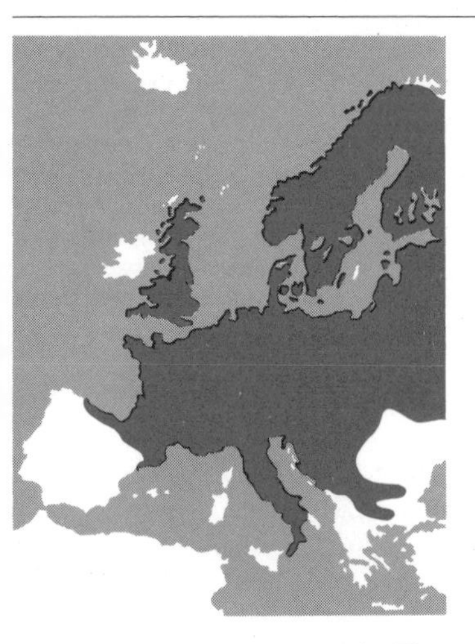

nostrils directed more upwards. The 30 teeth have red-brown tips.

Distribution: Found throughout Europe, except Ireland, the major part of the Iberian Peninsula, a large part of the Balkans and several islands. This species occurs in all parts of Scandinavia except along the coasts of the Arctic Ocean. To the east it extends through northern Asia to the Pacific.

Habitat: Mainly in lakes, ponds and rivers—providing the water is fairly clean but it also occurs some distance from water and even spreads into woodland.

Habits: The Water shrew is equally at home on land and in the water. It usually makes short excursions into the water, swimming along the bottom or in among vegetation in search of food; the latter consists of aquatic insects, larvae, worms, snails, small fishes, frogs and other animal food. The prey is paralysed or killed by a poisonous salivary gland secretion. Extensive subterranean tunnel systems are made; the tunnels are oval in cross-section and measure about 1·0 × 1·5 cm ($\frac{3}{8} \times \frac{5}{8}$ in.). In April–May a compact nest of moss and leaves is built in the ground, usually just below the surface. Here the female gives birth to 3–8 young after a gestation period of about 24 days. The young weigh about 1 g ($\frac{1}{30}$ oz) at birth and about 4 g ($\frac{1}{7}$ oz) at an age of 4 weeks. They are suckled for up to 37 days. Each female normally produces 2–3 litters in the course of a summer. The young become sexually mature in the following summer. Only the young survive through the winter.

Southern water shrew
Neomys anomalus

Identification: A little smaller than the preceding species and very similar to it in colour, but the belly is always white. Body-length 64–88 mm ($2\frac{1}{2}$–$3\frac{1}{2}$ in.); tail 42–64 mm ($1\frac{5}{8}$–$2\frac{1}{2}$ in.); hind-foot 14–18 mm ($\frac{1}{2}$–$\frac{3}{8}$ in.); weight 5–16 g ($\frac{1}{6}$–$\frac{1}{2}$ oz). Ventral keel of long hairs on the tail lacking or poorly developed and the hairy fringes on the feet are shorter.

Distribution: Associated with mountain regions in central and western Europe, but in the eastern areas of its range it also occurs at lower altitudes.

Habitat and habits: Approximately the same as for the preceding species.

9 Pygmy white-toothed shrew
Suncus etruscus

Identification: One of the smallest of all mammals. Body-length 34–44 mm. ($1\frac{3}{8}$–$1\frac{3}{4}$ in.); tail relatively long, 24–29

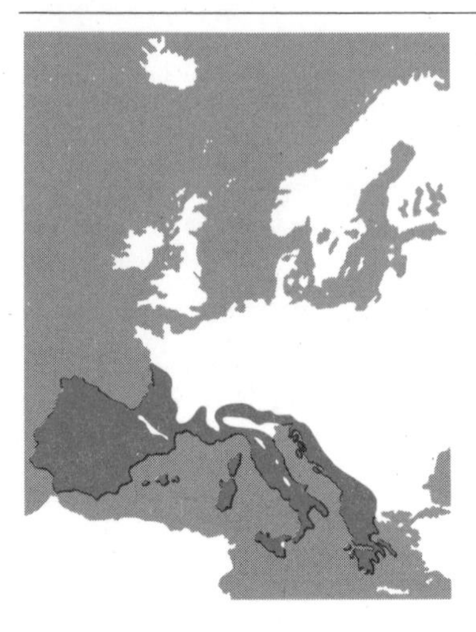

mm (1–1$\frac{1}{8}$ in.); weight only 1·5–2·0 g ($\frac{1}{20}$–$\frac{1}{15}$ oz). Back grey-brown with a reddish tinge, belly greyish and not sharply demarcated from the back.

Distribution: From Spain through South France and Italy to the southern part of the Balkans and farther east to the southern Soviet Union.

Habitat: Lives in fields, gardens, scrub and open woodland (cork oak and similar) in warm, dry stony areas. Frequently seen close to inhabited places and often in houses.

Habits: A delicate animal which can only survive in a subtropical climate; it does not tolerate temperatures below about 8° C (46° F). The food consists of all kinds of small animals, probably chiefly the smaller invertebrates.

10 Bicolour white-toothed shrew
Crocidura leucodon

Identification: A relatively large species. Body-length 64–87 mm (2$\frac{1}{2}$–3$\frac{3}{8}$ in); tail less than half the body-length, 28–39 mm (1$\frac{1}{16}$–1$\frac{1}{2}$ in.); weight 6–12 g ($\frac{1}{5}$–$\frac{2}{5}$ oz). Back slate-grey to brownish-black, contrasting sharply with the whitish belly. Teeth: see fig. 10a. Note that the canine tooth in the upper jaw is relatively small, and lower than the foremost cusp of the succeeding, very large cheek tooth.

Distribution: Eastern and central Europe. Absent from the Iberian Peninsula, western France, the British Isles, Denmark, Scandinavia, Finland and the Baltic countries.

Habitat: Lives in more open places such as the outskirts of woods, gardens, stone walls, heathland, fields and dry meadows; sometimes close to outbuildings which it often enters.

Habits: Diet consists of worms, snails, insects, other arthropods and carrion. Like the other species of *Crocidura* it builds a large nest of grass, lined with feathers and down. Apart from a

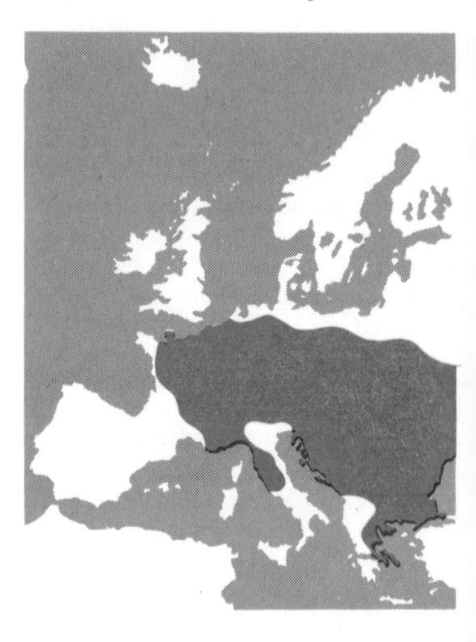

small, circular entrance-hole the nest is enclosed. If disturbed when the young are only a few days old, the mother carries them away in her mouth to a new hiding-place. When the young are larger—about 10 days old—they move off in a highly original but effective manner when danger threatens: one of the young seizes the skin at the root of the mother's tail in its teeth, and the other young grips the skin of the next one in front in the same way, thus forming a chain. The mother then leads the whole caravan away to safety. They evidently hold on very firmly because if the mother is lifted off the ground the chain remains unbroken and they are all lifted up together.

11 Lesser white-toothed shrew
Crocidura suaveolens

Identification: A relatively small species. Body-length 53–82 mm (2–3¼ in.); tail 24–44 mm (1–1¾ in.) (a good half the length of the body); weight 3–11 g ($\frac{1}{10}$–$\frac{1}{3}$ oz). Back very variable in colour: greyish, brownish or yellowish-red, not sharply demarcated from the paler belly.

Distribution: From south-east Europe through central Europe to parts of France and north-west Spain. It is thus absent from the major part of western and northern Europe, but has isolated populations in France and is also found in the Isles of Scilly.

Habitat and habits: Mainland populations usually in more open country and sometimes in deciduous woodland. Habits approximately the same as those of the preceding and following species but comparatively little is known about its behaviour, probably because this species is easily overlooked.

12 Common white-toothed shrew
Crocidura russula

Identification: A relatively large species. Body-length 64–95 mm (2½–3¾ in.); tail 40–50 mm (1½–2 in.) (more than half the body-length); weight 6–12 g ($\frac{1}{5}$–$\frac{2}{5}$ oz). Back brownish-grey, belly whitish-grey, but no sharp line of demarcation between the upper and under sides. Teeth: see fig. 12b. Note that in the upper jaw the canine (the third tooth from the front) is relatively large and as long as the foremost cusp of the succeeding large cheek tooth.

Distribution: Throughout most of the Mediterranean area and parts of central Europe and southern Russia but absent from Denmark, Scandinavia and Finland. Found in the Channel Islands but not in Britain.

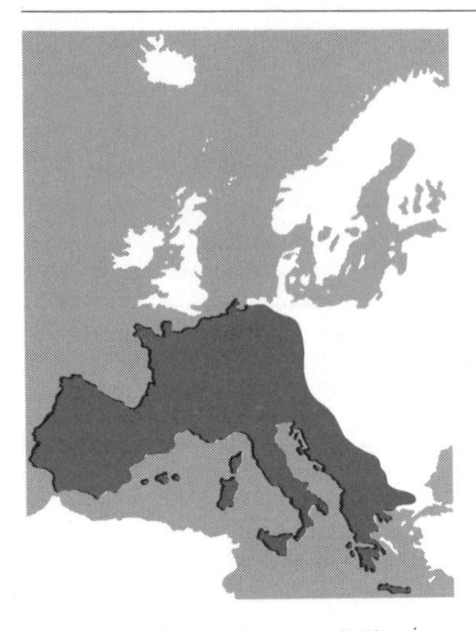

Habitat: As for *C. leucodon* (10).

Habits: More nocturnal than the species of *Sorex* and not so fast in its movements. Digs its own runs in the ground but also uses the underground systems of mice and moles. Out in the open this species breeds from March to August, but in outbuildings litters may be found at any time of the year. There are usually 3–6 young in each litter. The whole population is replaced in the course of a single year. See also under *C. leucodon* (10).

Moles

In Europe the family Talpidae includes the moles and desmans. These differ considerably in external characters, particularly in the development of the fore- and hind-limbs.

13 Pyrenean desman
Galemys pyrenaicus

Identification: A relatively large species with a long, rat-like tail and a long muzzle. The fur is very dense: metallic brownish on the back, silver-grey on the belly with a yellowish sheen. Body-length 110–135 mm ($4\frac{5}{16}$–$5\frac{1}{4}$ in.); tail longer than the body, 130–150 mm ($5\frac{1}{8}$–6 in.); hind-foot 31–38 mm ($1\frac{1}{4}$–$1\frac{1}{2}$ in.); muzzle-length about 20 mm ($\frac{3}{4}$ in.); weight 50–80 g ($1\frac{3}{4}$–$2\frac{3}{4}$ oz). The feet are flattened and fringed with stiff hairs; the hind-feet, in particular, are noticeably flattened and webbed. The innermost part of the tail is cylindrical and constricted at the root, thereafter it suddenly thickens and the distal part is lanceolate, laterally compressed and also has a fringe of hairs. In the thick part near the root of the tail there is a large scent gland which produces an oily secretion with a strong musky smell. There are 44 teeth of which the 2 front teeth are very large in relation to the remainder.

Distribution: The Pyrenees and the north-western part of the Iberian Peninsula. The related Russian desman, *Desmana moschata*, is found in the Soviet Union.

Habitat: Lives mainly along rivers and canals, near lakes and ponds where the water is clear and there is scrub and low vegetation, preferably with tangles of roots, decayed branches and leaves along the banks and on the bottom.

Habits: Desmans are adapted for aquatic life and obtain most of their food from the water. They run along the bottom, exploring with the very mobile snout and pulling worms and larvae from their burrows. They also feed on fish, frogs, snails and aquatic insects.

The system of burrows in the banks is partly dug by the desmans themselves but existing cavities are also used. The entrance is well below the water surface and a tunnel leads obliquely upwards and extends above the water level. There are no ventilation holes leading up to the soil surface. The nest is made under the roots of a bush or in a similar place well above the water level. It consists mainly of moss and is characterized by the dreadful smell and mass of food remains. There are probably two litters in the year. The above information applies mainly to the Russian desman. Little is known about the biology of the Pyrenean species but it appears to be less tied to water. In summer it can, for instance, be found in grass tussocks and under haystacks in fields.

14 Northern mole
Talpa europaea

Identification: A relatively large species. Body-length 115–150 mm ($4\frac{1}{2}$–6 in.); tail much shorter than the body, 20–34 mm ($\frac{3}{4}$–$1\frac{3}{8}$ in.); weight 65–120 g ($2\frac{1}{4}$–$4\frac{1}{4}$ oz). The fur stands out at right angles from the body and does not lie in any one direction, thus allowing the animal to move forwards and backwards with equal facility. The individual hairs are almost zigzag in form, so the pelt is very dense, silky-soft and impenetrable. The colour is slate-grey to black, with the belly usually somewhat paler. The fore-limbs (14a) are very powerful and adapted for digging. The very small eyes, about 1 mm (under $\frac{1}{16}$ in.) in diameter, are normally visible and have movable eyelids. There are 44 teeth.

Distribution: Found throughout most of Europe. In Scandinavia found only in the southern areas of Sweden and Finland. Absent from Ireland, the western and southern parts of the Iberian Peninsula, the southern Balkans and many of the large and small islands. The range extends eastwards through Asia to Japan.

Habitat: In cultivated fields and permanent pasture, gardens and woodland, but absent from marshy ground and areas of loose sand.

Habits: Moles spend the greater part of their lives underground in an extensive, branched system of burrows. Usually only one individual lives in each system. There are different types of burrows. The true hunting burrows may lie from 5 to 100 cm (2–40 in.) below the soil surface, but they are usually found at a depth of 10–20 cm (4–8 in.). The earth is dug with the fore-limbs and scraped backwards with the hind-limbs. When a suitable amount has been loosened, the animal turns and, using the fore-limbs, pushes the earth back through the burrow and

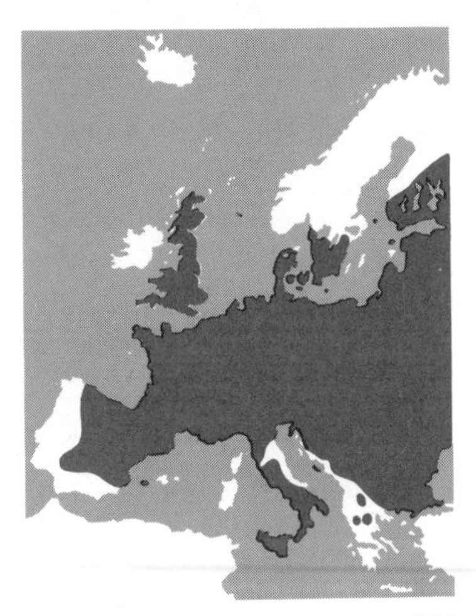

up through an oblique tunnel to the surface. Using the left and right forelimb alternately the soil is heaved up in a series of jerks to form the familiar mole-hills. A second type of burrow is formed when the mole digs just below the soil surface. Here it presses the loosened soil upwards so that it makes a wall at the surface. Straight tunnels very close to the surface are particularly noticeable in the spring when the male is searching for a mate.

The Northern mole traverses its system of burrows three times in the course of the 24 hours. Each inspection takes about 4½ hours, with an intervening rest period of about 3½ hours. During the rest periods the mole remains in a chamber formed in the central part of the tunnel system. Moles feed mainly on earthworms, beetles and fly larvae, snails and millipedes, but will also take small frogs and young mice. The daily food intake amounts to 40–50 g (1½–1¾ oz). The senses of touch and smell at short range are well developed, but hearing is poor. The eyes can probably only differentiate between light and darkness.

Gestation lasts about 4 weeks and during this period the female builds a large nest lined with leaves and grass; the nest chamber is usually at a considerable depth in the earth. The female produces 2–6 young, usually 4, in April or May to June. At birth these are naked and blind, but they are independent at an age of 4–5 weeks, when they weigh about 60 g (2 oz). Occasionally there is a second litter in August.

Moles are active throughout the winter. They then live deeper down in the ground and collect a store of earthworms, biting them so that they cannot move away but remain alive. Moles live for up to 3 years. They have many enemies. Foxes and badgers dig up the nests, and weasels hunt them in their own burrows. In summer many young moles are taken by owls and birds of prey. Finally, moles are also hunted by man. A field riddled with mole-hills is difficult to harvest, and in gardens and nurseries digging by moles may loosen young plants. Therefore farmers and horticulturalists regard the mole as a pest and it is either trapped or poisoned. In woodland, on the other hand, the mole's digging activities are regarded as beneficial as they help to aerate the soil.

Mediterranean mole
Talpa caeca

Identification: Very similar to the Northern mole (15), but a little smaller and the hairs on the lips, limbs and tail are paler. Sometimes described as the Blind mole as often no eye-opening is visible.

Distribution and habitat: Found throughout most of the Iberian Peninsula, apart from the north-eastern part, and also in isolated localities in north-west Italy, northern Yugoslavia, Macedonia, and eastwards along the south coast of the Black Sea to the Caucasus. A smaller type can be found in the same localities as the Northern mole close to the boundaries between the ranges of the two species.

Habits: Not essentially different from those of the Northern mole.

Bats

The bats which are distributed all over the world are classified in the order Chiroptera. There are about 875 species and among the mammals the order is second only to the rodents in number of species. The majority of bat species are found in the tropics, only one or two occur north of the Arctic Circle and none in the true polar regions. Most bats are relatively small and the dominant colours are dark brown and grey. The order is divided into two suborders: the larger bats or Megachiroptera and the smaller or Microchiroptera; the former are found exclusively in the tropics and contain species which are vegetarian, feeding mainly on soft fruits. The second suborder is by far the largest and it contains species which feed primarily on animal food, particularly insects. Nearly all bats are active only after dark.

Within the animal kingdom there are a number of species which can glide or soar for a limited distance but it is only the insects, birds and bats that are capable of true flight. In both birds and bats the fore-limbs are developed as wings. In the birds the supporting surface of the wing is formed by the rows of flight feathers positioned along the hind edge of the fore-limb. In the bats, on the other hand, the fore-limb itself provides the supporting structure. The usual musculature has almost disappeared, and the skin on the upper and lower surfaces is fused to form a very thin, flat wing-membrane. This is supported and kept taut during flight by the much elongated fingers. The wing-membrane, which is almost hairless, extends along the sides of the body to the ankles. There is usually a spur-shaped cartilaginous or bony structure, the calcar, which arises from the inner side of the ankle; the interfemoral membrane lies between the calcar and the tail. In some species there is an extension of skin from the calcar on the outer edge of the interfemoral membrane; this is known as the post-calcarial lobe. The first finger or thumb is always short and free and is equipped with a claw. The relatively slender hind-limbs can be turned in such a way that the feet face backwards. The feet have 5 toes, each with strong, curved claws. When at rest bats normally hang head downwards, suspended by the claws of the hind-limbs and with the wing-membranes either folded or enveloping the body.

In the Microchiroptera the eyes are very small. The external ears are well developed and very varied in form. The muzzle usually has a number of folds of skin around the nostrils. The dentition is much reduced and the molars usually have sharp cusps. The fur is extremely soft and the structure of the individual hairs is often very complex. Bats are closely dependent on the environmental temperature. As soon as a bat comes to rest—hanging upside down—it loses the ability to regulate its temperature upwards and gradually the body temperature will fall. This happens in diurnal sleep and also, of course, during hibernation; the period of hibernation depends entirely upon the local climatic conditions.

Bats rely to a great extent on their sense of hearing and the ears are extraordinarily well developed. Many species have a small lobe in front of the ear passage

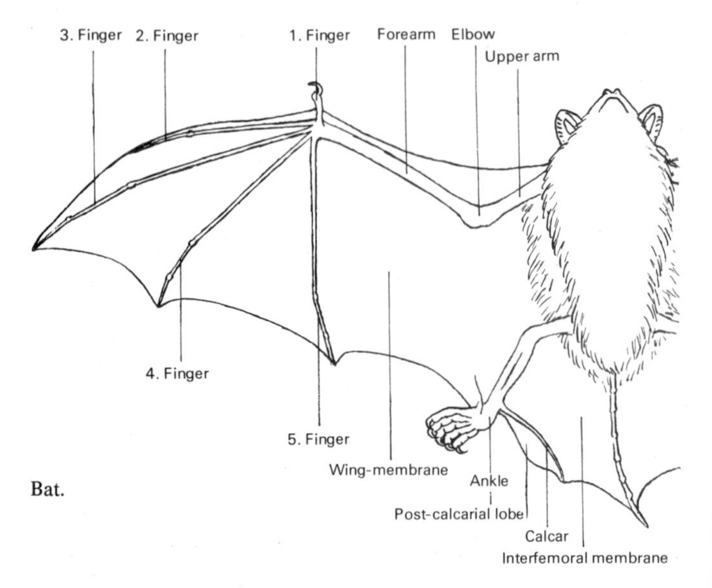

Bat.

known as the tragus. During flight, bats emit sound waves, either through the nose or the mouth; these are inaudible to the human ear and have a very high frequency varying according to the species. These sounds are reflected by objects in the immediate vicinity and when the information is received by the bat's ears, it provides an astonishingly detailed picture of the immediate surroundings. For instance, an insect flying by is detected at once and its position precisely located. In addition, bats produce other sounds which are audible to the human ear; these are mainly high-pitched squeaks.

Three families of bat are represented in the fauna of Europe: the horseshoe bats (Rhinolophidae), the vespertilionid bats (Vespertilionidae) and the free-tailed bats (Molossidae).

Horseshoe bats

Members of the horseshoe bat family or Rhinolophidae have characteristic leaf-like outgrowth of skin—known as nose leaves—around the muzzle. These are useful in distinguishing between the species. The lowermost is the so-called horseshoe, a fold of skin which covers the upper lip and forms the lower border of the nostrils. Above this there is a protruding, saddle-shaped outgrowth known as the sella which seen from the side has an upper and a lower protruding process. Finally, in the region of the forehead there is an erect, triangular fold of skin called the lancet. The ears have no tragus. When at rest the majority of species envelop their bodies in the wing-membrane.

15 Lesser horseshoe bat
Rhinolophus hipposideros

Identification: One of Europe's smallest bats. Body-length only 37–41 mm ($1\frac{1}{2}$–$1\frac{5}{8}$ in.); tail 24–30 mm (1–$1\frac{3}{16}$ in.); forearm 35–42·5 mm ($1\frac{3}{8}$–$1\frac{5}{8}$ in.); ear 15–17·5 mm ($\frac{5}{8}$–$\frac{3}{4}$ in.); wing-span about 20 cm (8 in.); weight 3–9 g ($\frac{1}{10}$–$\frac{1}{3}$ oz). Fur dense and soft, pale brown with a greyish sheen on the back, paler on the underparts. Seen from the front the sella is saddle-shaped, from the side almost square, both the upper and lower processes being poorly developed (15c). Horseshoe leaf relatively small.

Distribution: A southern species, although it extends farther north than any of the other European horseshoe bats. The northern limit runs from the British Isles—where it occurs chiefly in the southern half of the country but also in parts of northern England and in Ireland—through west and central Germany, Poland and the Carpathians to the Crimea. South of this line it is widely distributed, particularly in certain parts of southern Europe, extending eastwards to Asia Minor and Persia and southwards to Sudan and Eritrea. Absent from mountainous regions.

Habitat: Originally associated with caves but has adapted better than the other horseshoe bats to human habitations. In Europe it now occurs mainly in large buildings such as castles, churches and ruins and to a lesser extent in caves, mine shafts and rock crevices, and exceptionally in hollow trees.

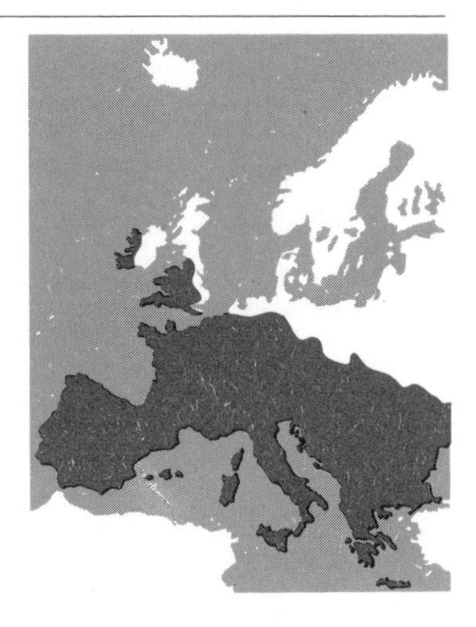

Habits: In the resting position, it normally hangs free and not in a crevice, head downwards and suspended by the feet. The body is enveloped in the wing-membrane (15b) and thus there is considerable water loss through evaporation. Cool sites are used for roosts in summer, but in winter they require well-sheltered places, such as caves and mine shafts. In summer the females form large or small nursing colonies which may contain up to 100 individuals. They spend the winter alone or in small groups, and are sometimes found in the underground part of the same building where the colony has lived during the summer. On the other hand, they may fly to different quarters for the winter and distances of up to 50 km (30 miles) have been recorded. Hibernation lasts from September to May, the optimum temperature being in the range 4–7° C (41–45° F). Mating takes place during autumn, winter or spring, but the embryos do not start to develop until the spring. The period of gestation is about 75 days. Occasionally there are 2 young in a litter although normally there is only one. The young are born from late June to July. The eyes open about 1 week after birth and the fur is fully developed after 2 weeks. Sexual maturity is reached at an age of $1–1\frac{1}{2}$ years.

This species only flies out when it is completely dark and then only when the weather is warm. The flight is low and fluttering, with rapid quivering wing-beats. Diet includes moths, flies, small beetles and spiders.

16 Mediterranean horseshoe bat
Rhinolophus euryale

Identification: Body-length 43–58 mm ($1\frac{3}{4}–2\frac{1}{4}$ in.); tail 22–30 mm ($\frac{7}{8}–1\frac{1}{8}$ in.); forearm 45–53 mm ($1\frac{3}{4}–2\frac{1}{8}$ in.); ear 18·5–24 mm ($\frac{3}{4}–1$ in.); weight 8–14 g ($\frac{1}{4}–\frac{1}{2}$ oz). Smoky-grey to red-brown on the back, whitish-grey or pale brownish on the underparts. Upper process of sella very pointed and turned upwards (16a). Lancet triangular. Horseshoe leaf small.

Distribution: The Mediterranean countries northwards to central France, Austria and the Caucasus. Also in northern Africa and parts of western Asia. Absent from the British Isles.

Habitat: A typical cave-dwelling species, which often occurs in large colonies, both in summer and winter.

Habits: In southern Europe the summer colonies may number thousands of individuals of both sexes. They overwinter singly or gregariously, hanging free from the roofs of caves but, unlike the other horseshoe bats, the wing-membrane does not envelop the body. Hibernation is not particularly deep and the animals fly when disturbed. They fly out to feed at twilight. A single young is born in late June.

Blasius' horseshoe bat
Rhinolophus blasii

Identification: Very similar to the preceding species. Body-length 44–51 mm ($1\frac{3}{4}–2$ in.); tail 24–25 mm (1 in.); forearm 43·5–49 mm ($1\frac{3}{4}–1\frac{15}{16}$ in.); ear 16·5–21 mm ($\frac{5}{8}–\frac{7}{8}$ in.). Colour relatively pale: back yellow-brown with a greyish sheen, underparts whitish tinged with grey-brown. Sella similar to that of the preceding species but upper process shorter (16b). Lancet concave at the sides. Horseshoe leaf broad.

Distribution: Found in countries around the eastern Mediterranean. Italy, Albania, Greece and in Cyprus and other islands. It also occurs in northern Africa and parts of western Asia. Absent from the British Isles.

Habitat: As in the preceding species this bat lives only in caves in very warm areas.

Habits: Little information available but probably much the same as those of the preceding species.

Mehely's horseshoe bat
Rhinolophus mehelyi

Identification: Very similar to the two preceding species but a little larger. Body-length 55–64 mm ($2\frac{1}{8}$–$2\frac{1}{2}$ in.); tail 24–32 mm (1–$1\frac{1}{4}$ in.); forearm 50–55 mm (2–$2\frac{1}{8}$ in.); ear 18–23 mm ($\frac{3}{4}$–$\frac{7}{8}$ in.). Upper process of sella not particularly long but very narrow at the tip and turned down.

Distribution: Spain, southern France, Sardinia, Rumania and Transcaucasia. Absent from the British Isles.

Habitat and habits: As for the Mediterranean horseshoe bat (16).

17 Greater horseshoe bat
Rhinolophus ferrumequinum

Identification: The largest of the European horseshoe bats. Body-length 56–69 mm ($2\frac{1}{4}$–$2\frac{3}{4}$ in.); tail 30–43 mm ($1\frac{1}{4}$–$1\frac{5}{8}$ in.); forearm 53–60·5 mm ($2\frac{1}{8}$–$2\frac{3}{8}$ in.); ear 20·5–26 mm ($\frac{3}{4}$–1 in.); wing-span 35–40 cm ($13\frac{1}{2}$–$15\frac{3}{4}$ in.); weight 15–27 g ($\frac{1}{2}$–1 oz.). Back grey-brown to red-brown, underparts paler. Seen from the front the sella is distinctly narrow in the middle; the upper process is broad and rounded, and looks short when seen from the side (17a). Horseshoe leaf relatively large and broad.

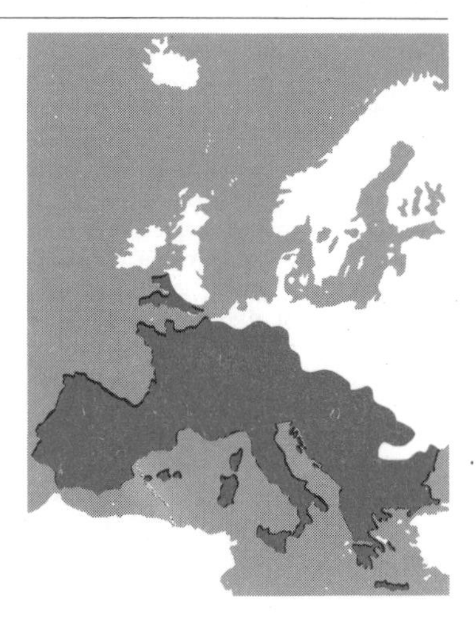

Distribution: A southern species, with its northern limit in Europe extending from the southern half of Britain—absent from Scotland and Ireland—through southern Holland, Germany, Austria and Hungary eastwards to parts of Russia, Japan and China. It is also found in north Africa and many areas of western Asia. There are several records of this species being found during winter in localities north of the main range of distribution.

Habitat: Lives mainly in subterranean cavities, particularly limestone caves and mine shafts; also in rock crevices and ruined buildings. In some parts of its range lofts and attics are frequently occupied in warmer weather.

Habits: During the summer they spend the day hanging from the ceiling in the warmer parts of the caves or other sites. Nursing colonies of females with their young may contain up to 500 individuals. The adult males mostly live solitarily during the summer but one or two may be found in the nursing colonies. This species moves to its summer quarters during March–April. The nursing colonies assemble during May and the young are born in June–July. Usually there is only a single young but occasionally there are two. After about 2 months the young are adult-size. During August the summer colonies break up and by October they have settled in damp caves at temperatures above 10° C (50° F), where they hibernate. The winter quarters are normally not far from the summer quarters, up to 50 km (30 miles), but there are records of individuals which have moved up to 150 km (90 miles) away. Like the Lesser horseshoe bat this species only flies out to hunt for food after dark and then only on warm, dry almost windless nights. The flight is slow—somewhat reminiscent of a butterfly—only a few feet above the ground. Diet consists of various night-flying insects.

Vespertilionid bats

Most of the European bats belong to the family Vespertilionidae in which the species lack prominent outgrowths of skin around the muzzle, but have a tragus, which varies in form, in front of the ear passage. The tail is completely or almost completely embraced by the interfemoral membrane. When at rest the wing-membrane is folded along the sides of the body.

18 Long-eared bat
Plecotus auritus

Identification: Easily recognizable by the long ears, 32–43 mm ($1\frac{1}{4}$–$1\frac{3}{4}$ in.), which have 20–25 transverse folds and are wide at the base, the inner edges meeting on the top of the head. Tragus long and pointed, 12–18 mm ($\frac{1}{2}$–$\frac{3}{4}$ in.)—about half as long as the ear—and up to 5·5 mm ($\frac{3}{16}$ in.) wide. Muzzle pointed with deep-set nostrils on the upper side. Wings broad; wing-membrane extends to base of outer toe. Thumb over 6 mm ($\frac{1}{4}$ in.) long. No post-calcarial lobe. Tail relatively long, 34–50 mm ($1\frac{3}{8}$–2 in.), with only the extreme tip free from interfemoral membrane. Body-length 41–51 mm ($1\frac{5}{8}$–2 in.); forearm 35–41 mm ($1\frac{3}{8}$–$1\frac{5}{8}$ in.); wing-span 24·5–25·5 cm ($9\frac{3}{4}$ in.); weight 5–10 g ($\frac{1}{6}$–$\frac{1}{3}$ oz). Colour of back varies from dark brown to pale brown.

Distribution: Widely distributed in Europe and Asia, from the British Isles in the west to China and Japan in the east. Reaches latitude 63° N in Scandinavia and Finland. Only found locally in the mountain regions of southern Europe. Also occurs in North Africa.

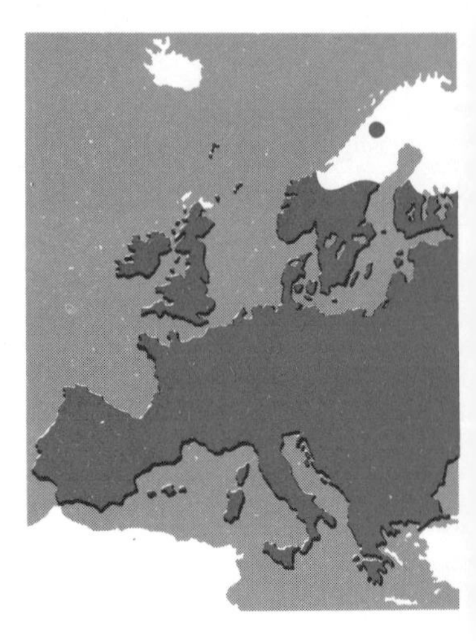

Habitat: One of the commonest and best-known bats in northern and central Europe. It appears originally to have been associated primarily with woodland but has adapted well to living in association with human habitations. The summer colonies are often found under the roofs of buildings and in hollow trees. The winter quarters are more frequently in cellars, outhouses, caves and mine shafts—in the latter two sites always close to the entrance.

Habits: Summer colonies of females contain 5–20 individuals. One or two young are born in June–July and are able to fly at an age of 1 month. Hibernation lasts from September to May in northern Europe, from October to March–April in central Europe, but it is interrupted at intervals. The bats move around in mild weather and may even merge into the open for brief sorties. Individuals usually hibernate singly, hanging from the roof or walls or in crevices. The large ears are folded back along the sides of the body while the bat is sleeping but the large tragus still projects (18b). The winter quarters are seldom far from the summer quarters—usually less than 30–40 km (18–24 miles).

Activity starts at twilight, the bats emerging before it is completely dark. They remain active throughout the night, gliding around tree-tops and houses, and flying low over vegetation. The flight is slow and fluttering, and this species can hover while catching insects in the air or taking them direct from foliage. Sometimes they land on tree-stumps or on the ground, continuing to forage at this level where they display greater agility than other bats. Larger prey is often consumed at fixed feeding-places. During flight the ears are directed forwards and the body is held somewhat obliquely in the air, giving a characteristic flight silhouette. Diet consists of moths, butterflies, mayflies, gnats, craneflies, cockchafers and other insects.

Grey long-eared bat
Plecotus austriacus

Identification: Very similar to the preceding species but a little larger. Ear-length 33–41 mm ($1\frac{1}{4}$–$1\frac{5}{8}$ in.); tragus 16–20·5 mm ($\frac{5}{8}$–$\frac{3}{4}$ in.) long, over 5·5 mm ($\frac{1}{4}$ in.) broad and with black pigment at the tip. Body-length 41·5–58 mm ($1\frac{5}{8}$–$2\frac{1}{4}$ in.); tail 46–57 mm ($1\frac{3}{4}$–$2\frac{1}{4}$ in.). Wings and ears less transparent than in the preceding species. Back ash-grey or grey-brown without yellowish or reddish tinge.

Distribution: From Spain and North Africa to central Asia. In Europe northwards to southern England, northern France, south Holland, central Germany and Poland and the Ukrainian Carpathians. In these more northerly areas it is only found in warm localities.

Habitat and habits: From the biological viewpoint not very different from the preceding species, but it is more warmth-loving and is more closely associated with human habitations.

19 Schreibers' bat
Miniopterus schreibersii

Identification: A medium-sized species with very long, pointed wings. Ears remarkably short—10–13·5 mm ($\frac{3}{8}$–$\frac{1}{2}$ in.)—projecting only slightly beyond the fur. Tragus short, 5–7·2 mm ($\frac{3}{16}$–$\frac{1}{4}$ in.) long, uniform in width, tip rounded and turned inwards. There is a clear demarcation between the erect hair on the head and the silky hair on the body. Body-length 52–60 mm (2–$2\frac{3}{8}$ in.); tail 50–60 mm (2–$2\frac{3}{8}$ in.); forearm 44–47 mm ($1\frac{3}{4}$ in.); wing-span 29·5–30·5 cm (12 in.); weight 10·5–17 g ($\frac{1}{3}$–$\frac{1}{2}$ oz). Back grey-brown, underparts whitish ash-grey. Wing-membrane extends to the ankles; inter-femoral membrane embraces the whole tail.

Distribution: A warmth-loving species found in Europe mainly in the Mediterranean countries and northwards into central France, southern Germany, Hungary and Poland. Occurs also in large areas of Africa, most of southern Asia—and eastwards to China, New Guinea and Northern Australia.

Habitat: Typically in caves, also in mine shafts, both in summer and winter. Occasionally found in cellars and lofts during summer.

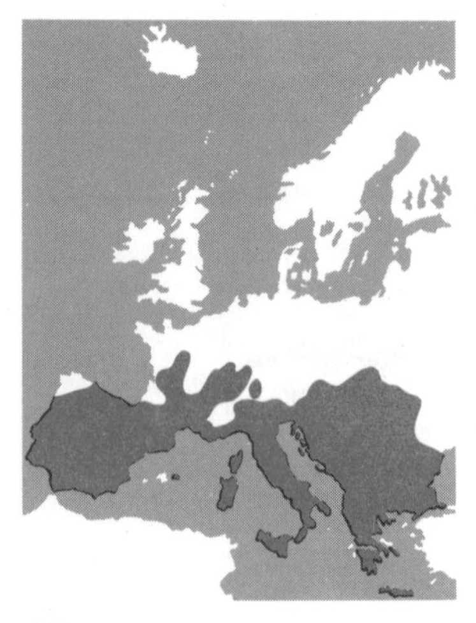

Habits: Large summer colonies may contain thousands of females with young. Colonies have been found in central Asia with up to 40,000 individuals. These bats often hang in large groups high up under the roofs of caves. Mating takes place in autumn before hibernation. The eggs are fertilized immediately but further embryonic development is delayed during hibernation. The single young is born in May–July. In the great majority of species the young bat is reared by its own mother but in tropical areas Schreibers' bat apparently adopts a communal system of rearing. Observations on huge colonies of this species in western India suggest that each female gives birth to a single young but the young are crowded together in large clusters and as each female lands she suckles the two youngsters that happen to be nearest to her.

Large numbers are found sharing the same winter quarters. Hibernation is not particularly deep. This species often flies great distances, travelling up to several hundred miles, between the summer and winter quarters. It is one of the strongest fliers among the bats, flying high and fast over open ground. It emerges shortly after sunset.

20 Barbastelle
Barbastella barbastellus

Identification: A medium-sized species with very broad ears which meet on top of the head. Outer edge of ear concave, with a small knob-like projection a little above centre. Ear-length 12–19·5 mm ($\frac{1}{2}$–$\frac{3}{4}$ in.); tragus 9–11 mm, broad at the base but tapering rapidly to a point. Muzzle broad and short, nostrils on the upper side. Wings short, wing-membrane extending to

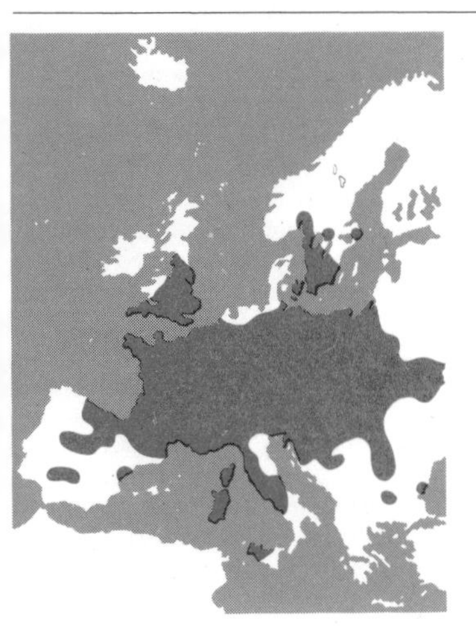

the base of the outer toe. Post-calcarial lobe present. Tip of tail free from interfemoral membrane. Body-length 44–58 mm ($1\frac{3}{4}$–$2\frac{1}{4}$ in.); tail 41–54 mm ($1\frac{5}{8}$–$2\frac{1}{8}$ in.); forearm 37–42 mm ($1\frac{1}{2}$–$1\frac{5}{8}$ in.); wing-span 26·5–27·5 cm ($10\frac{1}{2}$ in.); weight 6·5–14 g ($\frac{1}{5}$–$\frac{1}{2}$ oz). Colour brownish-black with a whitish-grey sheen on the back, grey-brown on the underparts. Fur dense and long, present on the inner parts of wing- and interfemoral membranes.

Distribution: Widespread in central and western Europe but absent from parts of Spain and Italy, and almost entirely from the Balkans. Locally in England and Wales but not in Scotland and Ireland. Present in parts of southern Sweden but absent from Finland. Also occurs in the Caucasus, Crimea and Ukraine.

Habitat: Found in summer in crevices in rocks and walls and in hollow trees, also behind loose bark and the boarding of outhouses. Spends the winter in cellars and caves, more rarely in mine shafts.

Habits: In western Europe the summer colonies contain 20–50 females, each of which usually produces 2 young in July. In England this species is often solitary but sometimes very small colonies are found, usually in woodland areas. Winter colonies with up to 2,000 individuals have been found in central Europe but the size of the colonies is very variable and some individuals hibernate on their own. Hibernation lasts from November to March–April, but is not very deep. This species is not particularly sensitive to low temperatures and is often found quite close to the entrance of a cave in winter. In some parts of their range these bats return faithfully to the same winter quarters, often covering distances of 200–300 km (120–185 miles) from their summer quarters.

Activity starts immediately after sunset and even takes place in cold and rainy weather. They fly along the edges and paths of woodland, in orchards and around houses. The flight is at a height of 1·5–5 m (4–16 ft), rather heavy and fluttering, changing direction frequently, but there are occasional bursts of speed. A second period of activity takes place in the early morning. They feed on small insects.

21 Pipistrelle
Pipistrellus pipistrellus

Identification: Smallest European bat. Body-length 33–52 mm ($1\frac{1}{4}$–2 in.); tail relatively short, 26–33 mm (1–$1\frac{1}{4}$ in.); forearm 27–34 mm (1–$1\frac{1}{4}$ in.); ear broad and short, 10–12 mm ($\frac{3}{8}$ in.); tragus short, 4–7·5 mm ($\frac{1}{8}$–$\frac{1}{4}$ in), rounded at the tip; wing-span 18–21 cm ($7\frac{1}{2}$ in.); weight 3·8–7 g ($\frac{1}{7}$–$\frac{1}{4}$ oz). Back uniformly dark but colour varies from dark brown to reddish, paler on the underparts. Post-calcarial lobe small and narrow (21c). Fur extends to both sides of wing-membrane. The latter extends to base of outer toe. Last tail vertebra free from interfemoral membrane. Fifth finger less than 43 mm ($1\frac{3}{4}$ in.) in length. Thumb relatively short (21b). Teeth: see fig. 21d. The size and position of the front teeth are significant in distinguishing species of *Pipistrellus*.

Distribution: Widespread in most of Europe, including the British Isles, and eastwards through Asia to Japan and Korea. In Scandinavia found only in southern Sweden and the extreme south of Norway.

Habitat: In open country and built-up areas, including large towns. In the mountains of central Europe it reaches altitudes of up to 2,000 m (6,500 ft).

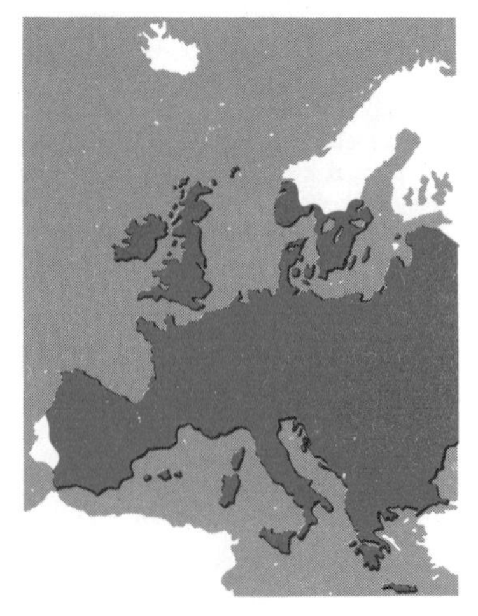

Often found in buildings in summer, living in cracks in masonry, behind loose boards, under roof tiles, behind pictures and in similar confined spaces. It can also be found in hollow trees, under loose bark and in nest-boxes. The winter quarters are similar but some are found hibernating in caves and mines.

Habits: In summer colonies of females may contain 50–200 individuals, while the males are usually found living solitarily or in small groups. In winter large numbers of this species may be found hibernating together, squeezed into cracks and crevices but never hanging free. The young are born between late May and July, occasionally in early August. In England there is usually only a single young but in other parts of the range the litter-size is two. The young become independent in about 2 months. Alleged records of migrations of over 1,000 km (620 miles) are not supported by satisfactory evidence. European populations apparently overwinter in the areas where they spent the summer. Hibernation is normally from late October to March but may be interrupted. When temperatures fall below −10° C (14° F) these bats emerge and seek warmer quarters, e.g. in houses.

This species starts to hunt at sunset and there are several periods of activity during the night. More than one individual can often be seen flying together over the same small area. The flight is fast and fluttering, with numerous quick wing-beats and changes of direction, usually at heights of up to 6 m (20 ft). They feed on flies, moths and other small insects. This species will also fly in broad daylight.

22 Nathusius' pipistrelle
Pipistrellus nathusii

Identification: Very similar to the preceding species but a little larger. Body-length 44–48 mm ($1\frac{3}{4}$–$1\frac{7}{8}$ in.); tail 34–40 mm (about $1\frac{1}{2}$ in.); forearm 32–36·5 mm ($1\frac{1}{4}$–$1\frac{3}{8}$ in.); ear 12–14·5 mm ($\frac{1}{2}$ in.); tragus 6–8 mm ($\frac{1}{4}$ in.), more convex on rear surface; wing-span 23·7–24·5 cm ($9\frac{1}{2}$ in.); weight 6–12 g ($\frac{1}{5}$–$\frac{2}{5}$ oz). Upper side of interfemoral membrane densely haired. Paler than the preceding species: back dark grey-brown, underparts paler. Wings broader, the fifth finger longer than 43 mm ($1\frac{5}{8}$ in.) and the thumb also longer (22a). Teeth: see fig. 22b.

Distribution: An eastern species which only occurs sporadically in western Europe, but more regularly east of a line from Hamburg to the Rhône delta. Not recorded in Britain and Finland, and in Scandinavia only from the most southerly part of Sweden. The range extends eastwards to the Caucasus and Black Sea.

Habitat and habits: In general similar to the preceding species.

23 Kuhl's pipistrelle
Pipistrellus kuhli

Identification: Very similar to the Pipistrelle (21). Body-length 40–47 mm ($1\frac{1}{2}$–$1\frac{3}{4}$ in.); tail 30–40 mm ($1\frac{1}{4}$–$1\frac{5}{8}$ in.); forearm 31–37 mm ($1\frac{1}{4}$–$1\frac{1}{2}$ in.); ear short, 12–13 mm ($\frac{1}{2}$ in.); tragus also short, 5·5–6 mm ($\frac{1}{4}$ in.); wing-span 22·6–23 cm (9 in.). Paler than the Pipistrelle (21): back dull greyish-yellow or yellowish, underparts white with a yellowish sheen. There is a sharply defined white edge to the wing-membrane between the fifth finger and the hind-foot. Lengths of calcar, fifth finger and thumb as in 21. The two species differ considerably in their dentition (compare figs. 23a and 21d).

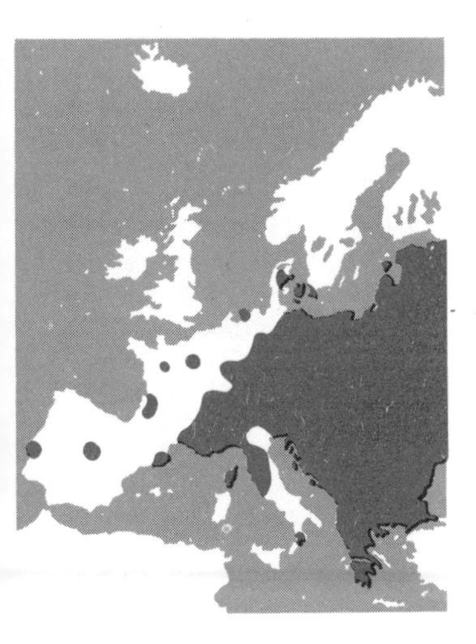

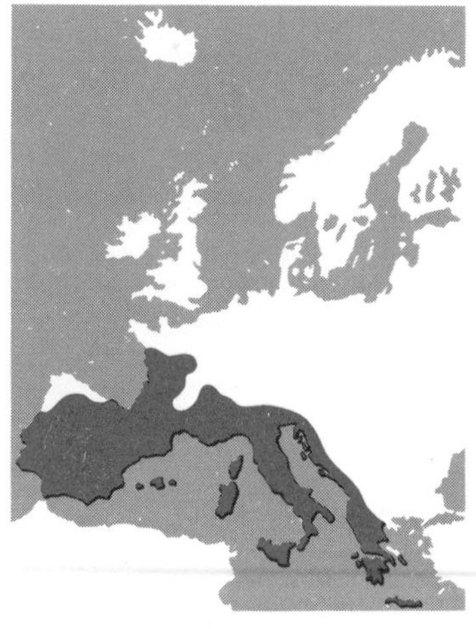

Distribution: Western Europe from Spain, France and Switzerland eastwards to Asia Minor, Persia and Afghanistan. Also found in Africa and some parts of southern Asia. Absent from the British Isles.

Habitat and habits: One of the commonest bats in southern Europe. Warmth-loving and therefore found mainly in the lowlands. Often seen flying around houses in towns and villages. The flight is fast, following a straight course, often skimming low over obstacles.

24 Savi's pipistrelle
Pipistrellus savii

Identification: Larger than the Pipistrelle (21). Body-length 43–48 mm ($1\frac{3}{4}$–$1\frac{7}{8}$ in.); tail 34–39 mm ($1\frac{3}{8}$–$1\frac{1}{2}$ in.); forearm 31–40 mm ($1\frac{1}{4}$–$1\frac{5}{8}$ in.); ear longer, 10–17 mm ($\frac{3}{8}$–$\frac{5}{8}$ in.), and less pointed; tragus 4·5–6 mm ($\frac{1}{4}$ in.)—less than half the ear-length—broad at centre, tapering rapidly towards tip and with a small toothlike outgrowth on the outer edge. Thumb short. Wing-membrane extends to base of outer toe as in the Pipistrelle (21). Wing-span 22·3–23·5 cm (9 in.). Fur dense and long. The hairs are of two colours, those on the back being brownish-black at the base and red-brown at the tips, while on the underparts they are pale brownish at the tips. Teeth: see fig. 24a.

Distribution: South Europe from central and eastern Spain through south France and Italy to Greece and the Caucasus. Northwards to Switzerland and Bavaria. Also in north Africa and parts of Asia.

Habitat: Appears to have been associated originally with forests in mountain areas. It reaches altitudes of up to 2,600 m (8,500 ft) in the Alps but is also found at low altitudes in the Mediterranean countries, where it enters buildings to some extent. Often in hollow trees, rock crevices and other cavities in summer. In mountain regions there is probably a vertical migration between summer and winter quarters.

Habits: Little data available. Small summer colonies of females have been recorded with only 10–30 individuals. Two young, rarely only one, are born at the end of June. A few males have also been found in these colonies during late summer.

Activity starts after sunset and continues throughout the night. This species flies along the edges of woods, in clearings and over alpine meadows. The flight is slower and more leisurely than in the Pipistrelle (21).

25 Daubenton's or Water bat
Myotis daubentoni

Identification: A small species. Body-length 41–51 mm ($1\frac{1}{2}$–2 in.); tail 30–39 mm ($1\frac{1}{8}$–$1\frac{1}{2}$ in.); forearm 35–41 mm ($1\frac{3}{8}$–$1\frac{5}{8}$ in.); ear (25a) medium-sized, 11–17 mm ($\frac{1}{2}$–$\frac{5}{8}$ in.), with a slight indentation on upper half of outer edge. Tragus 5–8 mm (about $\frac{1}{4}$ in.), barely half the ear-length and very narrow towards the tip. Calcar extends two-thirds along edge of interfemoral membrane. Wing-membrane ends at middle of hind-foot (25a). Wing-span 21–25 cm ($8\frac{1}{4}$ in.); weight 5–11 g ($\frac{1}{6}$–$\frac{1}{3}$ oz). Back dark grey-brown with a reddish sheen, sharply demarcated from greyish-white underparts.

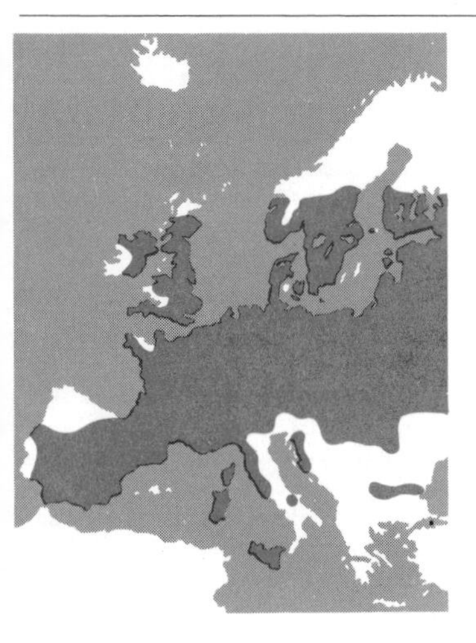

Distribution: Found in most parts of northern Europe, including Britain and Ireland, and southwards to Spain, Italy and Rumania. In Scandinavia and Finland it reaches latitude 63–64° N. The range extends eastwards through Asia to China and Japan.

Habitat: Associated with areas of woodland with ponds and lakes. Not common in dry areas, such as southern Europe where it occurs only locally. During the summer it lives in hollow trees, lofts, crevices in buildings and similar places, and in winter in caves, mine shafts and cellars.

Habits: Summer colonies are found, varying in size, in which the sexes are not always segregated. The single young is born in June–July. Mating takes place in the winter quarters, during the autumn, winter or spring. In central Europe this species undertakes considerable migrations—often more than 100 km (60 miles)—from summer to winter quarters. The most favourable temperature in the winter quarters appears to be 5–9° C (41–48° F). Hibernation lasts from October to April, but is frequently interrupted. This species is usually solitary during winter but sometimes several are found close together either in crevices or hanging free.

Activity starts immediately after sunset and continues through the night. This species flies low over the water, often at a height of only 30 cm (1 ft), repeatedly dipping down to the water surface. A large number of individuals will often forage together; this species also swims well. Prey includes flies, moths and other flying insects.

26 Pond bat
Myotis dasycneme

Identification: A relatively large species. Body-length 57–61 mm ($2\frac{3}{8}$ in.); tail 46–51 mm ($1\frac{3}{4}$–2 in.); forearm 43–49 mm ($1\frac{5}{8}$–$1\frac{7}{8}$ in.); ear (26a)

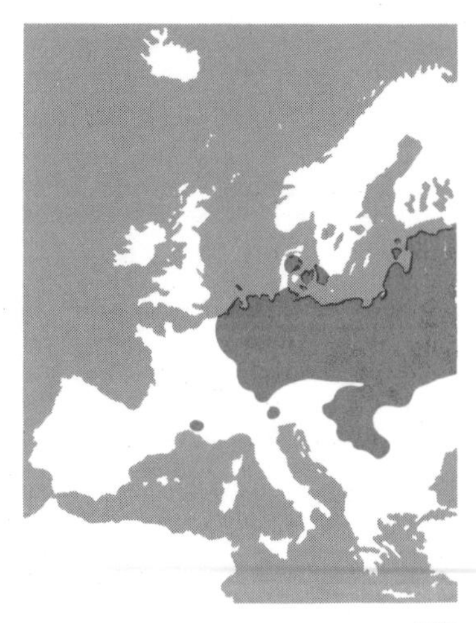

relatively small, 14–17 mm ($\frac{5}{8}$ in.), with 4–6 transverse folds, tapering somewhat towards tip. Tragus 6·6–8·5 mm ($\frac{1}{4}$ in.), barely half as long as the ear. Calcar as in the preceding species. Wing-membrane ends at ankle (26a). Wing-span 20–30 cm (8–12 in.); weight 15–17 g (about $\frac{1}{2}$ oz). Back dark grey-brown, underparts greyish-white.

Distribution: The breeding range extends from northern France, Belgium and Holland eastwards through Germany, Poland and the Soviet Union to Siberia. They also overwinter in Czechoslovakia, Austria, Switzerland, Hungary, Rumania, north Italy and northern Yugoslavia but do not breed in these countries. Absent from the British Isles.

Habitat: Associated with ponds, lakes and marshes in wooded areas in low-lying country. Summer colonies are found in houses, church towers, hollow trees and similar places. In winter this species is mainly solitary but groups may be found in caves, mine shafts, rock crevices and cellars.

Habits: In Europe the summer colonies of females contain only 5–20 individuals, but in the Soviet Union colonies have been found with several hundred individuals. The males live solitarily or in small groups and are often found with other species. Almost nothing is known of the breeding biology of this species. It usually hangs free from the roof or behind rafters. The winter quarters are often a considerable distance from the summer quarters. Marked individuals have moved up to 330 km (190 miles), having travelled in a southerly or south-westerly direction. Hibernation takes place from October to April.

This species emerges late, flying only when it is completely dark. It hunts over water, flying low and sweeping above the surface without any sudden twists and turns.

27 Long-fingered bat
Myotis capaccinii

Identification: A medium-sized species. Body-length 47–53 mm ($1\frac{7}{8}$–$2\frac{1}{8}$ in.); tail 35–38 mm ($1\frac{3}{8}$ in.); ear (27a) 14–16 mm ($\frac{5}{8}$ in.), comparatively narrow with indentation on outer edge. Tragus long and narrow, 6·5–7·5 mm ($\frac{1}{4}$ in.). Calcar as in the preceding two species. Wing-membrane ends at ankle (27a). Feet strikingly large. Upper side of interfemoral membrane noticeably hairy. Last tail vertebra free from interfemoral membrane. Back grey-brown, underparts whitish-grey.

Distribution: Mediterranean area, including eastern Spain, south France, south Switzerland, Italy, Balkans and eastwards to western Asia. Also in North Africa.

Habitat: Only in warm areas in low-lying country. Colonies are found in caves and mine shafts in winter and summer. Sometimes found sharing the same quarters with other species.

Habits: Little information available but probably similar to those of Daubenton's and the Pond bat; it is known to hunt over water.

28 Large mouse-eared bat
Myotis myotis

Identification: One of the largest European bats. Body-length 68–80 mm ($2\frac{5}{8}$–$3\frac{1}{8}$ in.); tail 48–60 mm ($\frac{7}{8}$–$2\frac{3}{8}$ in.); forearm 59–67 mm ($2\frac{3}{8}$–$2\frac{5}{8}$ in.); ear 24–31 mm (1–$1\frac{1}{4}$ in.), broad with 7–8 transverse folds. Tragus 11–15 mm (about $\frac{1}{2}$ in.), scarcely half the ear-length. Wing-span 35–40 cm ($13\frac{1}{2}$–15 in.); weight 18–40 g ($\frac{3}{5}$–$1\frac{1}{3}$ oz). Wings grey-brown and relatively broad; wing-membrane ends at base of outer toe. Last tail vertebra free from interfemoral membrane. Back grey-brownish to pale reddish-grey; underparts whitish-grey.

Distribution: In many parts of Europe including Portugal, Spain, France, southern England, Germany, Switzerland, Italy, Sardinia, Hungary, Poland and Rumania, and eastwards to Persia and China.

Habitat: Mainly in the lowlands but occurs at altitudes of up to 1,900 m (6,200 ft) in the Alps. Found in built-up areas as well as in the countryside, often associated with arable land

not far from water. Summer colonies are found in lofts and also in caves. In winter this species lives in caves, mine shafts and cellars.

Habits: A highly gregarious species. The summer colonies of females may contain 50–3,000 individuals, which usually hang free from the roof, sometimes in clusters and even in several layers. In cold weather they may move into crevices and cracks. This species is active from the end of March to the middle of September or later if the mean temperature is above 10° C (50° F). The females live in the summer colonies from April to September; the males rejoin the females towards the end of June and mixed colonies are found until the following March. There appears to be a strict segregation of the sexes during the early summer. The winter quarters are often some distance from the summer quarters—up to 260 km (160 miles) in central Europe. Only the northern populations migrate to the south-west in autumn. The atmosphere is humid in the winter quarters and the bats usually hang free, either solitarily or in clusters.

Activity starts when it is completely dark. The flight is very low, 5–8 m (16–25 ft) above the ground, slow and direct. In mild weather this species can be seen on the continent flying over roads, across open ground, in gardens, parks and along the edges of woods. Diet consists mainly of insects—preferably the larger species. Mating takes place late in autumn or spring. The period of gestation is 50–70 days, depending upon the temperature. The female produces 1, or exceptionally 2, young in July. The maximum life-span is 15 years.

29 Whiskered bat
Myotis mystacinus

Identification: One of the smallest European bats. Body-length 38–50 mm ($1\frac{1}{2}$–2 in.); tail 30–40 mm ($1\frac{1}{4}$–$1\frac{1}{2}$ in.); forearm 32–38 mm ($1\frac{1}{4}$–$1\frac{1}{2}$ in.); ear (29a) 12–14·5 mm ($\frac{1}{2}$ in.), with 4–6 transverse folds, its outer edge slightly rounded. Tragus very narrow, longer than half the ear-length, 7–9 mm ($\frac{3}{8}$ in.). Calcar extends halfway along the edge of interfemoral membrane. Wing-membrane extends to base of outer toe (29a). Wingspan 20–22·5 cm (8–9 in.). Fur dense and silky, extending on to both surfaces of wing and interfemoral membrane.

Colour variable: back grey-brown or red-brown to almost brownish-black—often with golden sheen of pale-tipped hairs—the underparts paler. Muzzle very hairy, hence the name. Weight 4·5–13·5 g ($\frac{1}{5}$–$\frac{1}{2}$ oz).

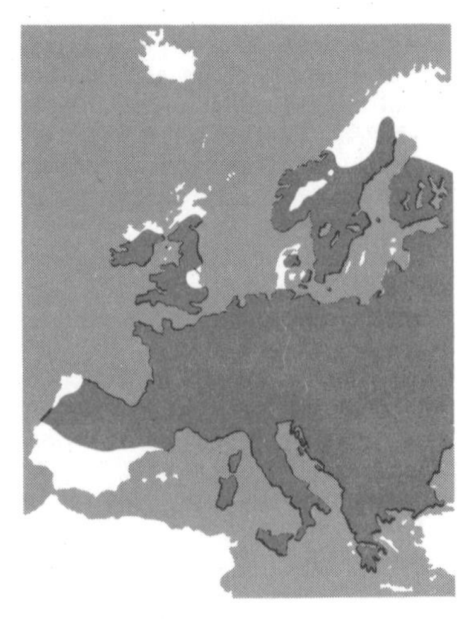

Distribution: Widespread over the whole of Europe, including most of Britain and Ireland. Absent from the southern half of the Iberian Peninsula, and the northern parts of Scandinavia and Finland. Found also in large areas of Asia, eastwards to China and Japan.

Habitat: Prefers areas with woodland and small lakes. Summer colonies live in hollow trees and in buildings. The winter is spent in caves, mine shafts and cellars.

Habits: During the summer the two sexes are segregated, the males living solitarily and the females in colonies of 10–50 individuals. The young are born from mid-June to mid-July. Records refer to single births, to complete independence at the age of 1 month and to sexual maturity in the second year for females. Hibernation starts in October and lasts until April–May but is often interrupted in mild weather. The sleeping bats hang free on the walls or move into crevices, and are found either solitarily or in small groups. This is a very hardy species which tolerates temperatures below freezing-point and a high air humidity. The quarters are usually only a short distance from the summer quarters—less than 50 km (30 miles). These bats often return to the same winter quarters for several years running. The annual mortality is about 20 per cent.

A very active species which often flies in daylight. Activity starts very early in the evening and probably continues throughout the night, with short resting periods. They fly at a height of 2–6 m (6–20 ft) along the edges of woods, around houses and over water. The flight follows a regular and comparatively narrow beat. By comparison with the Pipistrelle (21), another species which also flies by day, the flight of the Whiskered bat is slower and steadier.

30 Natterer's bat
Myotis nattereri

Identification: A little larger than the preceding species. Body-length 42–50 mm ($1\frac{5}{8}$–2 in.); tail 32–43 mm ($1\frac{1}{4}$–$1\frac{5}{8}$ in.); forearm 35–43 mm ($1\frac{3}{8}$–$1\frac{5}{8}$ in.). Edge of interfemoral membrane has a fringe of stiff, curved hairs on either side of tail (30a). Calcar extends about half-way along edge of interfemoral membrane, and has a slight bend. Wing-membrane extends to base of outer toe. Ear (30a) long and narrow, 14–20 mm ($\frac{1}{2}$–$\frac{3}{4}$ in.), with 5–6 transverse folds, outer edge slightly notched. Tragus also long and narrow, about two-thirds the length of the ear. Wing-span 24·5–25 cm (10 in.); weight 5–10 g ($\frac{1}{6}$–$\frac{1}{3}$ oz). Back grey-brown with a reddish sheen, underparts whitish.

Distribution: Widespread in Europe but only common in central Europe and in England, Wales and Ireland—rarely reported in Scotland. Extends into south Sweden and south Norway but in the northern and southern parts of the range it is less common. Also occurs in parts of Russia and eastwards to Japan.

Habitat: In woodlands and parks, including built-up areas. In mountain regions it extends up to the tree-limit. Summer colonies are formed in hollow trees and lofts, sometimes also in nest-boxes. The winter is usually spent in caves, mine shafts and cellars, sometimes in hollow trees.

Habits: The sexes are segregated and the males live solitarily. The evening

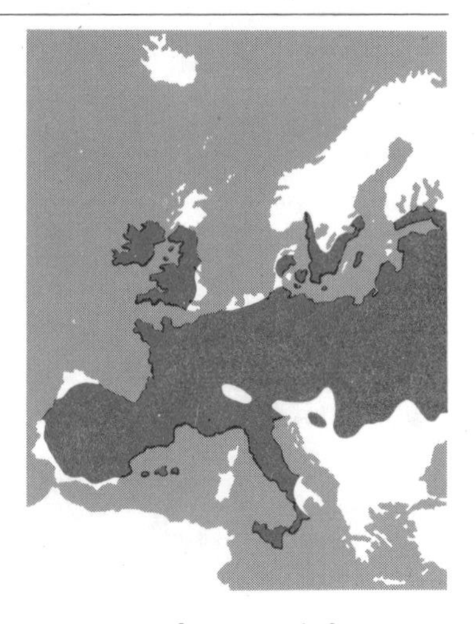

emergence often starts before sunset and activity may continue at intervals throughout the night. This species flies low along woodland rides and the edges of woods but also flies quite high around trees, searching for insects. The flight is slow and steady without sharp changes in direction. Diet consists mainly of various flying insects but not all prey is caught in flight.

31 Geoffroy's bat
Myotis emarginatus

Identification: Body-length 44–50 mm ($1\frac{3}{4}$–2 in.); tail 40–43 mm ($1\frac{5}{8}$ in.); forearm 36–40 mm ($1\frac{1}{2}$ in.); ear (31a) long, 16–17 mm ($\frac{5}{8}$ in.), with 6–8 transverse folds, outer edge with a deep notch. Tragus 9·5–11 mm ($\frac{3}{8}$ in.), thus more than half the ear-length. Wing-membrane includes base of outer toe (31a). Extreme tip of tail free from interfemoral membrane. Wing-span about 23 cm (9 in.); weight 6–9 g ($\frac{1}{5}$–$\frac{3}{10}$ oz). Back yellow-brown or

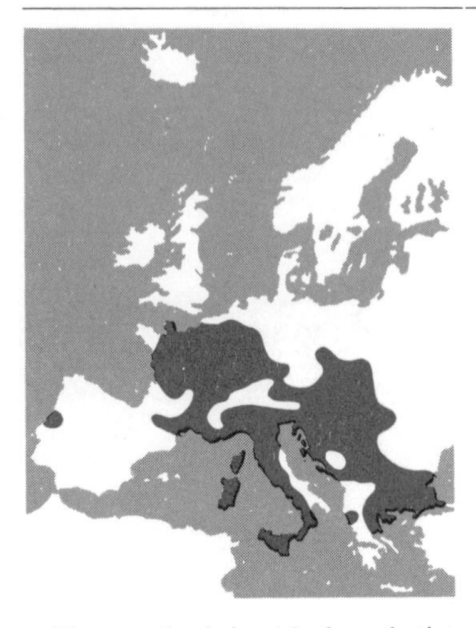

red-brown, the hairs tricoloured: tip reddish, middle part greyish and base brownish-black. Underparts whitish with a reddish sheen. Wings and ears pale brownish.

Distribution: Central and southern Europe. The northern limit runs through south Holland, west and south Germany, Czechoslovakia, south Poland and the Carpathians. The species also lives in parts of Asia. Absent from the British Isles.

Habitat: Mainly in warmer areas in the lowlands. In summer usually in lofts, hollow trees and behind loose bark. In winter in the depths of large caves, mine shafts and cellars, where the temperature is relatively constant.

Habits: Little information available. Observations include summer colonies of females with up to 500 individuals, often in company with horseshoe bats, the birth of a single young in June–July and the ability to fly at an age of 4 weeks. Winter quarters are often at a considerable distance from the summer quarters and in some cases are farther north. Emergence starts late in the evening; flight is at a very low height, often over water.

32 Bechstein's bat
Myotis bechsteini

Identification: A little larger than the three preceding species. Body-length 46–53 mm ($1\frac{3}{4}$–2 in.); tail 34–44 mm ($1\frac{3}{8}$–$1\frac{3}{4}$ in.); forearm 40–45 mm ($1\frac{1}{2}$–$1\frac{3}{4}$ in.); ear long and broad, 23–26 mm (1 in.), with 8–10 transverse folds, outer edges slightly concave. The ears project 10–13 mm ($\frac{3}{8}$–$\frac{1}{2}$ in.) beyond the tip of the muzzle when they are turned forwards. Tragus barely half as long as the ear, 10·5–14 mm (about $\frac{1}{2}$ in.). Wing-membrane extends to base of outer toe. Last vertebra free from interfemoral membrane. Wing-span about 22·5 cm (9 in.); weight 7–11 g ($\frac{1}{4}$–$\frac{1}{3}$ oz). Back

grey-brown with a reddish sheen, the hairs very long and bicoloured, dark brown at the base. Underparts whitish-grey, wings, ears and muzzle dark brown.

Distribution: Central and southern Europe northwards to south England, south Sweden and Lithuania, but only common in certain areas of central Europe. The southern limit runs through Bulgaria, Yugoslavia, central Italy and north Spain. Extensive finds of subfossil skeletal remains indicate that this species was formerly common in many parts of Europe.

Habitat: Associated with trees, especially in mountain regions. In summer colonies are found in hollow trees, nest-boxes, behind loose bark and in similar places. In winter this species is also found in hollow trees but usually hibernates in caves and mine shafts. Roosts have also been recorded under roofs of houses.

Habits: Little information available. Summer colonies of females have been recorded on the continent with up to 20 individuals. These colonies do not appear to assemble in the same place for several years in succession. A single young, born in June, has been recorded in central Europe. In winter quarters it is said to hang free, rarely in crevices, either in small groups or solitarily. There are few records of this species in England; it is said to fly rather slowly, near the ground, either just after sunset or later in the evening.

33 Serotine
Eptesicus serotinus

Identification: One of the largest European bats. Body-length 62–80

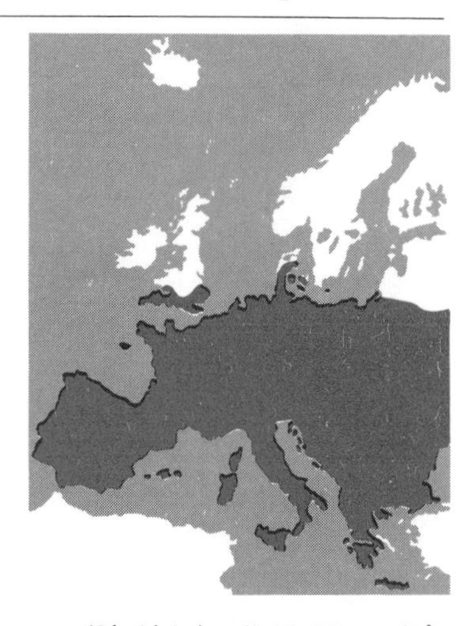

mm ($2\frac{1}{2}$–$3\frac{1}{8}$ in.); tail 46–57 mm ($1\frac{3}{4}$–$2\frac{1}{4}$ in.) with the last 7–8 mm ($\frac{1}{4}$ in.) extending beyond the interfemoral membrane; forearm 45–57 mm ($1\frac{3}{4}$–$2\frac{1}{4}$ in.); ear (33a) 18–23 mm ($\frac{3}{4}$–$\frac{7}{8}$ in.) and tragus short, 7–11 mm ($\frac{1}{4}$–$\frac{3}{8}$ in.), with a rounded tip. Post-calcarial lobe small. Wing-membrane extends to the base of the outer toe. Wing-span 34–35 cm ($13\frac{1}{2}$ in.); weight 15–35 g ($\frac{1}{2}$–$1\frac{1}{16}$ oz). Fur long, hair brown to reddish-brown on back but paler at the tips, underparts yellowish-brown. Sparsely haired on interfemoral membrane.

Distribution: The whole of central and southern Europe northwards to south England and Denmark. Absent from Scandinavia and Finland. Extends eastwards through the Soviet Union to eastern Asia.

Habitat: This is one of the commonest bats in south and central Europe, living in built-up areas as well as in parkland and wooded country. Summer colonies are usually found in buildings, including the roofs of houses and also in hollow trees. In winter this species usually lives in cellars, outhouses and hollow trees, more rarely in caves and mine shafts.

Habits: Summer colonies of females contain 10–30 individuals which hang free, while the males live solitarily. The young (1 or 2) are born in June–July. Hibernation starts in late September and may continue until the end of April. This species often travels long distances—up to 300 km (185 miles)—between the summer and winter quarters.

Emergence takes place shortly after sunset and there appear to be intermittent periods of activity through the night. They circle over gardens, parks and roads at a height of 3–20 m (10–60 ft). The flight is rather laboured and fluttering, with sudden darts at flying insects. Wing-beats can sometimes be heard quite distinctly. This species is evidently sensitive to weather conditions and does not fly on cold, rainy nights.

34 Particoloured bat
Vespertilio murinus

Identification: A medium-sized species. Body-length 55–63 mm ($2\frac{1}{8}$–$2\frac{1}{2}$ in.); tail 40–45 mm ($1\frac{1}{2}$–$1\frac{3}{4}$ in.) with the last vertebra extending beyond the interfemoral membrane; forearm 41–48 mm ($1\frac{5}{8}$–$1\frac{7}{8}$ in.); ear (34a) short and broad, 11·5–18 mm ($\frac{1}{2}$–$\frac{3}{4}$ in.). Tragus short and broad, 5–8 mm (about $\frac{1}{4}$ in.), shaped like a bean and turning in at the tip. Outer edge of ear continues down to mouth, forming a small pocket. Post-calcarial lobe narrow. Wing-membrane extends to base of outer toe. Wing-span 26·5–28 cm (11 in.); weight 11–14 g ($\frac{1}{3}$–$\frac{1}{2}$ oz). Back brownish-black, hair tips whitish or brownish-white, giving a silvery sheen; whitish on throat and in area between hind-limbs, rest of underparts pale yellowish-grey or brownish-white.

Distribution: In Europe mainly in the central and eastern parts extending west to certain areas of France and Germany, northwards to the southern districts of Norway, Sweden and Finland, and eastwards through the Soviet Union to Japan. Absent from the Iberian Peninsula and large areas of south and west Europe. Only rarely recorded in Britain.

Habitat: Found in towns and woodland in low-lying country, also in mountains in central and eastern Europe, but is nowhere common. During the summer it has been found in lofts, rock crevices, behind cracks in masonry and in hollow trees. The win-

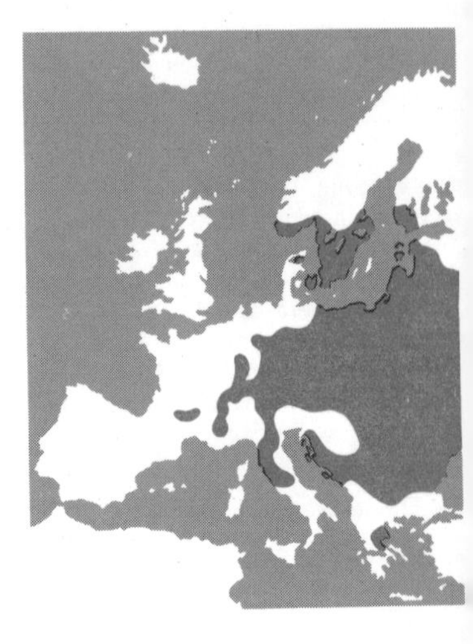

er is spent in cellars, caves and mine hafts.

Iabits: In Europe summer colonies of emales have been found with 40–50 ndividuals—often with other species resent in the colony—the males living solitarily or in separate colonies; he young (1 or 2) are born in June–uly. In Russia, however, segregation f the sexes in the summer colonies is ot so strict, a few living among the emales, and the number of young is nore frequently 2. It is likely that coniderable distances are covered when noving to the winter quarters. Hiberation cannot be very deep as this pecies is seen on the wing in winter vhen the weather is mild.

Emergence takes place after sunset nd activity ceases at sunrise. The bats ly high, around tree-tops and buildngs but they can be identified by their haracteristic calls which carry some listance.

35 Northern bat
Eptesicus nilssoni

dentification: Smaller than the two receding species. Body-length 48–54 nm ($1\frac{7}{8}$–$2\frac{1}{8}$ in.); tail 38–47 mm ($1\frac{1}{2}$–$\frac{7}{8}$ in.). Last tail vertebra projects 2–3 nm ($\frac{1}{8}$ in.) beyond interfemoral memrane. Size and form of ear and tragus s in the Particoloured bat, but outer dge of ear ends just behind mouth 35a). Forearm 37–43 mm ($1\frac{1}{4}$–$1\frac{3}{4}$ n.), post-calcarial lobe narrow. Wingnembrane extends to base of outer oe. Wing-span 24–27 cm ($9\frac{1}{2}$–$10\frac{3}{4}$ n.); weight 8–14·5 g ($\frac{1}{4}$–$\frac{1}{2}$ oz). Fur ong, hairs brownish-black on back ut yellow at the tips, giving a golden heen, underparts pale yellowishrown.

Distribution: Eastern and northern Europe. Found in the northern parts of the Soviet Union, Poland and Czechoslovakia and eastwards through north Asia to the Pacific Ocean. Also in Scandinavia and Finland where it reaches latitude 69° N, which is farther north than any other bat. In central Europe it occurs locally in mountain regions, but is absent from west and south Europe. Not recorded in the British Isles.

Habitat: Associated with forest and cultivated land. In Scandinavia extends up to the tree-limit and to an altitude of 2,000 m (6,500 ft) in the Alps and Carpathians. In summer the day is spent in lofts, rock crevices and hollow trees. In similar places during the winter, often in wooden buildings, but the bats move around if the first site becomes too cold.

Habits: Summer colonies of females are small, with up to 50 individuals. The sexes are segregated and the males

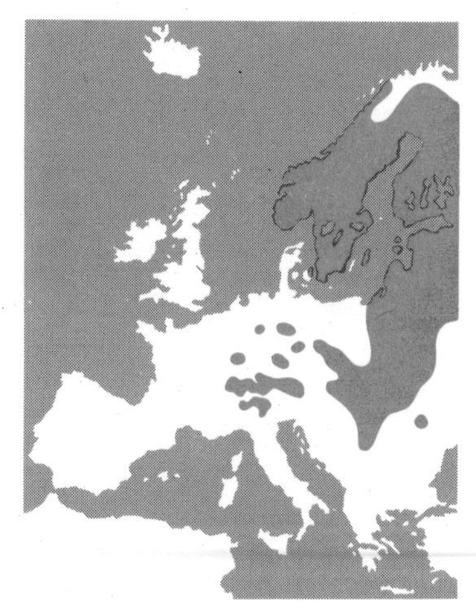

live solitarily. One or two young are born between mid-June and mid-July. This species emerges very early in the evening. In Scandinavia where the summer nights are light, it flies throughout the night in forests, but only over open ground as the nights become darker. During early spring and autumn it may also be seen flying by day. The flight is fast at heights up to 10 m (12 ft). Hibernation lasts from September–October until April. These bats are very resistant to low winter temperatures. They probably migrate from summer to winter quarters but this has not been observed.

36 Lesser noctule
Nyctalus leisleri

Identification: Smaller than but otherwise similar to the Common noctule (37). Body-length 54–64 mm ($2\frac{1}{8}$–$2\frac{1}{2}$ in.); tail 39–44 mm ($1\frac{1}{2}$–$1\frac{3}{4}$ in.); forearm 40–46 mm ($1\frac{1}{2}$–$1\frac{7}{8}$ in.). Ears relatively a little longer, 14–16·5 mm ($\frac{5}{8}$ in.), and the tragus shorter and broader 6–8 mm ($\frac{1}{4}$ in.). Wing-span 26–30 cm (10–12 in.); weight 14–20 g ($\frac{1}{2}$–$\frac{2}{3}$ oz). The hairs are bicoloured: brownish-black at the base, red-brown at the tip.

Distribution: A relatively rare species with an extensive range: England, Ireland, France, Germany, Spain, Poland, Czechoslovakia, Rumania and eastwards into western Russia.

Habitat: Associated with forest regions. Summer quarters in hollow trees and under loose bark, wintering also in buildings.

Habits: Summer colonies contain 30–40 females which give birth to 1 or 2 young in June–July. Records exist of long migrations, for example fron Poland to Czechoslovakia, a distanc of about 400 km (250 miles). Emergence takes place shortly after sunse and the flight is very similar to that c the Common noctule.

Giant noctule
Nyctalus lasiopterus

Identification: The largest Europea bat. Apart from its greater size, th diagnostic characters of this specie are scarcely sufficient to distinguish i from the next species. Body-lengt 84–104 mm ($3\frac{3}{8}$–$4\frac{1}{8}$ in.); tail 55–6 mm ($2\frac{1}{4}$–$2\frac{5}{8}$ in.); ears 21–26 mm ($\frac{3}{4}$– in.), and longer than in the Commo noctule; tragus 7–8·5 mm ($\frac{1}{4}$ in.) weight 41–76 g ($1\frac{1}{3}$–$2\frac{1}{2}$ oz).

Distribution: Only a few records fron a number of countries: France, Italy Switzerland, Germany, Bulgaria an the Soviet Union.

Habitat and habits: Associated wit deciduous forests, but little is know of its habits.

37 Common noctule
Nyctalus noctula

Identification: A large, sturdily buil species with long, narrow wings and typically broad head with the ears se wide apart. Ears rounded, nearly a broad as their length, 16–21 mm ($\frac{5}{8}$– in.). Tragus very short and broad, 6–8·5 mm ($\frac{1}{4}$ in.), rounded and disc shaped at the tip. Post-calcarial lob broad (37b). Body-length 69–82 mr ($2\frac{3}{4}$–$3\frac{1}{4}$ in.); tail 41–59 mm ($1\frac{5}{8}$–2 in.); forearm 47–55 mm ($1\frac{3}{4}$–$2\frac{1}{8}$ in.) wing-span 37–46 cm (14–18 in.) weight 21–35 mm ($\frac{2}{3}$–$1\frac{1}{6}$ oz). Smoot reddish-brown fur on back, slightl

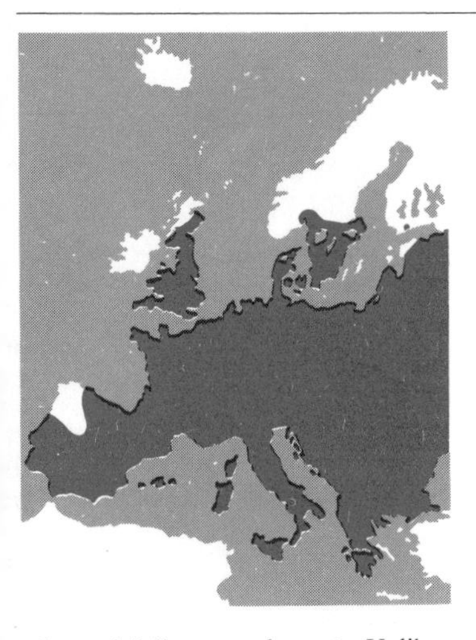

paler and duller on underparts. Unlike the previous species the hairs are all of one colour. Fur extends to both surfaces of wing- and interfemoral membrane and the long hairs on the underside of the upper arm are characteristic. Extreme tip of tail free from interfemoral membrane.

Distribution: Throughout most of Europe, and absent only from Ireland, north Scotland and Scandinavia north of about 60° N. Only recorded once in Finland.

Habitat: Deciduous and mixed woodland mainly in the lowlands, also in parks and gardens. In certain localities this species is very common. Summer and winter colonies are found in hollow trees and nest-boxes, often in holes made by woodpeckers. It is also found in wooden outhouses and, in winter, sometimes in stone buildings. Colonies never hibernate in caves although occasionally a solitary specimen is found there.

Habits: In summer the Common noctule spends the day in holes in trees. These holes are at a minimum height of about 6 m (20 ft) above the ground and the entrances become polished by the passage of bats going in and out. Roosts are changed frequently. The sexes are strictly segregated, the males living solitarily, except in the mating period. The female colonies contain 10–30 individuals. Mating takes place chiefly in the autumn and to a lesser extent in the spring; fertilization, however, only occurs some 5–10 days after the females emerge from hibernation. The gestation period is 70–75 days. The young are born about the middle of June, usually there are 2, more rarely only 1 and exceptionally 3. They are able to fly in August, when they are about the same size as the adults. Sexual maturity is reached at an age of 1 year in the females, 2 years in the males. For a short time in late summer and autumn colonies are formed with both sexes present, but these break up when the bats move to their winter quarters. It has been suggested that northern populations migrate between their summer and winter quarters. Although there are records of the recovery of marked individuals at a distance of over 2,000 km (1,250 miles), these are not sufficient to justify any theory of mass migration. Further work on populations wintering in Sweden has also thrown doubt on this theory. It is possible, however, that some of the younger bats migrate. Winter colonies may number thousands of individuals. True hibernation is prolonged in this species and it lasts from October–November to March–April. The winter quarters are not always frostproof and in very hard winters large numbers may die.

Common noctules start flying very early, often before sunset on dark days, and immediately after sunset on lighter days; they also fly during the day from time to time in the northern regions. The flight is often at a considerable height—5–25 m (16–80 ft) and sometimes up to 100 m (320 ft)—and around sunset this bat can be seen hunting for insects in company with swallows and martins; there are often two periods of activity during the night. This is one of the strongest fliers among the bats, flying fast and straight for a stretch, then twisting and turning rapidly, and diving down after prey. They even fly when the weather is bad. Shrill calls are often heard while the bats are hunting. Their diet consists of large insects such as moths and cockchafers.

Free-tailed bats

The family Molossidae is represented by one species in Europe, but there are many more species in the tropics and subtropics. The distinguishing feature of the free-tailed bats is the projection of the greater part of the tail beyond the interfemoral membrane.

38 European free-tailed bat
Tadarida teniotis

Identification: A very large species, only slightly smaller than the Giant noctule. Body-length 82–87 mm ($3\frac{1}{4}$–$3\frac{3}{8}$ in.); tail 46–57 mm ($1\frac{3}{4}$–$2\frac{1}{4}$ in.); forearm 57–63 mm ($2\frac{1}{4}$–$2\frac{1}{2}$ in.). Ears long, 27–31 mm (1–$1\frac{1}{4}$ in.), almost covering the broad, flat head; they face forwards and their inner edges meet.

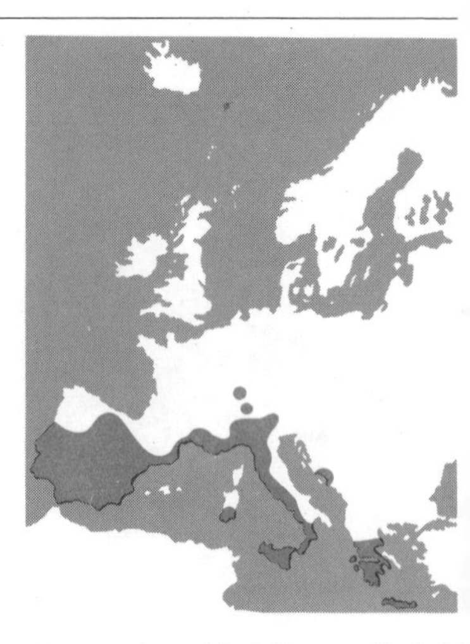

Tragus short, 6–6·5 mm ($\frac{1}{4}$ in.), broader than its length. Two-thirds of the tail projects beyond the interfemoral membrane. Long, narrow wings and short, sturdy legs.

Distribution: Southern Europe from Portugal and Spain through south France, south Switzerland and Italy to Yugoslavia and Greece, extending eastwards to parts of Russia and China. Absent from the British Isles.

Habitat and habits: Very little is known about this species which is rare within its range. It is present in rocky terrain, where small colonies roost in vertical fissures and under projecting ledges. Colonies have also been found in crevices of old buildings and in aqueducts. These bats emerge late—not until it is completely dark—and hunt throughout the night. The flight is fast and straight, fairly high, and resembles that of the swift. This species has only one young.

Lagomorphs

The lagomorphs (order Lagomorpha) were classified at one time among the rodents which they resemble in many ways. The order contains two families: the pikas or Ochotonidae, which are represented by a few species in north America and a larger number in Asia, and the Leporidae (hares and rabbits) which are found in almost every part of the world.

The Leporidae are mostly medium-sized animals with the following characteristics: the body is elongated and slim, the ears large, the limbs relatively long and slender, the hind-legs in some forms being particularly long. There are 5 toes on the fore-feet, 4 on the hind-feet. The tail is always very short. In the dentition the front teeth are separated from the cheek teeth by a section of jaw that is toothless. The canine teeth are completely absent. There is 1 incisor in each half of the lower jaw but the upper jaw has 2 in each half, the front incisor being much the larger of the two. All the incisors are long, narrow and curved; they are described as 'rootless' and the considerable wear to which they are subjected is replaced by continuous growth from a wide open pulp cavity. The front surface is much harder than the back, so that the latter wears away faster, with the result that the front of the tooth develops a sharp cutting edge and becomes chisel-shaped. The cheek teeth are also 'rootless' and their surfaces are thrown into projecting ridges of enamel. There are 6 cheek teeth in each half of the upper jaw, 5 in each half of the lower jaw. The chewing movements are mostly from side to side, in contrast to the rodents where the movements are mostly backwards and forwards. The eyes are large and in most forms are situated far out on the sides of the head, thus providing a wide field of vision. The sense of smell is well developed.

Lagomorphs feed exclusively on plant food and have a remarkably long gut. The caecum, a side-branch of the gut, is capable of holding ten times as much as the stomach. Food in the caecum is subjected to bacterial fermentation resulting in the production of faecal pellets that contain important vitamins. These moist faeces are different in appearance from the ordinary dry droppings that can be seen wherever hares and rabbits are found. The former are evacuated by the animal when it is at rest but are eaten again immediately. This process, known as refection, is comparable to rumination or chewing the cud.

Most lagomorphs live for only a few years but they are very prolific, producing several litters in a year. In many members of the Leporidae there is evidence of considerable cyclical fluctuations in population numbers.

Hares and rabbits

39 Rabbit

Oryctolagus cuniculus

Identification: Body-length 34–45 cm (13–17 in.); tail 4–8 cm ($1\frac{1}{2}$–3 in.), held vertically when running so that the white underside is visible. Hind-foot about the same length as the tail. Ears relatively short, usually without the pronounced black tips that are typical of the hare; when laid forward they do not reach the front of the muzzle. Back appears yellowish grey-brown, sides greyish, underparts white; completely black specimens occur. Weight 1·3–2·2 kg ($3\frac{1}{4}$–$5\frac{1}{2}$ lb).

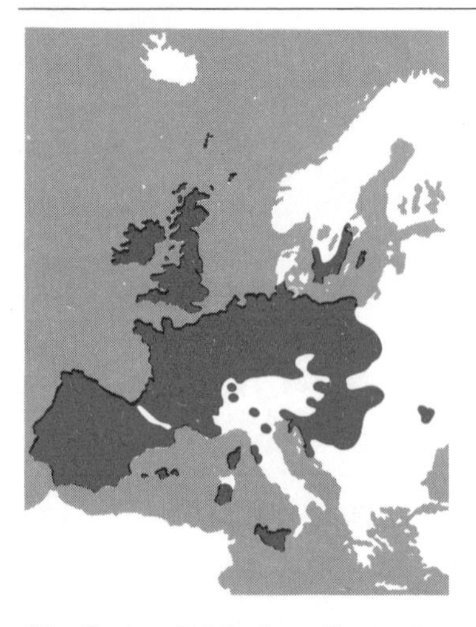

Distribution: Originally native to the Mediterranean area but has now spread over large areas of Europe, both by natural means and by the help of man. Also introduced to other parts of the world, such as Australia and New Zealand where it has been very successful.

Habitat: Associated with lighter soils in open country, among scrub, in rocky places and plantations.

Habits: Lives in colonies or warrens where a strict social hierarchy is maintained. The warrens are usually extensive, a complicated system of burrows being dug in the earth with several entrances, but they are also known to live completely or mainly above ground, for example in dense scrub or in rocky terrain. Rabbits spend most of the day in a nest chamber which is situated in the central part of the burrow complex, about 40–50 cm (15–20 in.) below the surface. They emerge in the late afternoon and are active throughout the night. Food consists of all kinds of plant matter, of which grass plays the most important role during the summer. Rabbits cause a great deal of damage to various field crops, vegetables and trees by nibbling shoots and gnawing bark. The daily food requirement is about 0·5 kg (1 lb) of vegetable matter. They move in very slow hops. When pursued they make rapid twists and turns, trying to reach a burrow as quickly as possible. Warning of danger is passed from one rabbit to another by thumping the ground with their hind-feet.

Under favourable conditions rabbits breed throughout the year, but in north temperate regions mainly during the spring and early summer. After a gestation period of 4 weeks the female gives birth to 3–12 young. These are born in a specially prepared breeding nest, lined with hay and fur from the mother's pelt, and usually made at the bottom of a specially dug burrow, which may be up to 1 m (3 ft) long; the entrance is blocked with earth when the mother leaves the young. The female usually becomes pregnant again within 12 hours of giving birth and may produce 3–5 litters in the year. The new-born young (39a) are naked and blind. They are suckled for about 4 weeks, after which they are able to look after themselves. Sexual maturity is achieved when 3–4 months old. Rabbits are hunted by many predatory mammals and birds, and at intervals populations are decimated by a virus disease known as myxomatosis. They are also subject to pest control in Britain and other countries. Rabbits make a high-pitched scream when injured or frightened. This is the species from which many domesticated races of rabbit have been developed.

40 Brown hare
Lepus capensis

Two forms of the Brown hare are found in Europe, and these are now considered to be one species. Until recently they were recognized as distinct species: *Lepus europaeus*, a slightly larger animal which is found in the north, and *Lepus capensis* (also known as the Cape hare) which occurs in parts of southern Europe and is also widely distributed in Africa.

Identification: Body-length of the northern form 48–68 cm (15½–27 in.); tail 7–11 cm (2½–4 in.); weight 2·5–6·5 kg (5½–14 in.); body-length of the southern form 40–54 cm (15–21 in.); tail 8–10 cm (3–4 in.); weight 1·5–2·5 kg (3¼–5½ oz). Tail is black on the upperside and held horizontally when running, leaving the white underside scarcely visible. Hind-legs and feet are noticeably long. Undersides of the feet are well-furred. Ears are long and distinctively marked with black tips; when laid forwards they extend about 3 cm (1¼ in.) beyond the front of the muzzle.

The colour of the coat is usually greyish-brown or reddish-brown—some specimens look paler than others—and due to the prominence of the black-tipped hairs, the overall appearance is more speckled, less uniform and grey than the rabbit. The southern form is said to have a whitish innerside of the leg which contrasts boldly with the reddish outerside of the thigh. In winter the coat is much the same colour as at other seasons, although, in general, the appearance is more drab.

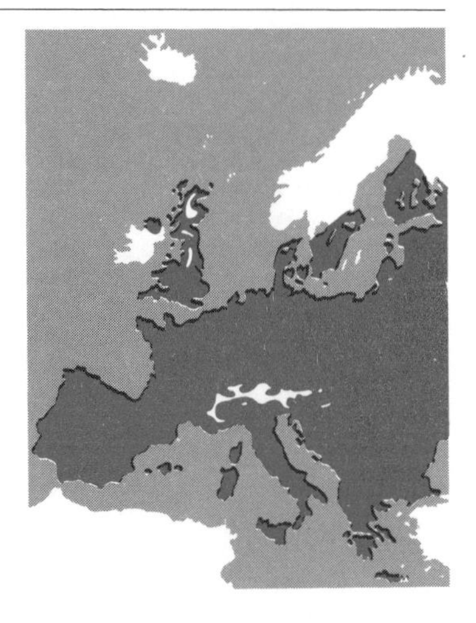

Distribution: The northern form is found in most parts of northern Europe, including Britain; in parts of France and Spain it overlaps with the southern form. It occurs eastwards to south-west Asia. Its distribution in Europe has been influenced by man owing to introductions in various areas for sporting purposes. For instance, until about 100 years ago, the Brown hare was totally absent from Scandinavia but it now extends into central Sweden and central Finland.

The southern form is found in the Iberian Peninsula, south of the River Ebro, the Balearic Islands and Sardinia, and also in many parts of Africa.

Habitat: Mainly in relatively flat open country, particularly on farmland, downs and heaths, but also present on rocky ground and in open woodland. It is not found in large blocks of continuous forest. The southern form lives in conifer forests on high ground.

Habits: Hares spend the day solitarily, lying in a 'form' which is a depression on the earth, usually well sheltered by vegetation. After sunset they come out to feed and are then more sociable. Direct observation after dark is difficult but as they are often seen feeding early in the morning, they are probably active from time to time throughout the hours of darkness.

Considerable distances may be covered between the form and the feeding areas, the animals following well-used routes known as hare paths. They rely particularly on their sense of hearing but smell is also well developed. The eyes are large and every movement is detected. Diet consists of grass, clover and other greenstuff; young cereal crops and root vegetables are also eaten, including garden produce. Damage is caused in forestry plantations and orchards, particularly in winter, when the hares attack the bark of trees. Young shoots of trees and shrubs are also liable to be eaten. Some of the food passes twice through the digestive system as the moist faecal pellets produced in the caecum are immediately eaten direct from the anus. Later the ordinary faeces are evacuated in the normal way; these are slightly flattened, yellow-brown, dry droppings, in which the coarser parts of the plant food can be clearly distinguished. Social displays which include chasing, boxing and acrobatic leaps, are popularly associated with the month of March but these displays may be seen much earlier if the weather is mild. Under these conditions mating may take place as early as January–February.

Several males can then be seen following a female and sometimes vigorous fights take place between rival bucks. Immediately after matings, the female releases an ovum thus ensuring effective fertilization. The gestation period is 42–44 days and the number of young per litter varies from 2 to 5; there are 3–4 litters per year. The newly born leverets (40a) are fully furred, their eyes are open and they are able to move around within a day. They are born on the ground, but, unlike the rabbit, not in a true nest; they are left to themselves except for short periods when visited by the doe for suckling.

The young are independent after 3 weeks, and reach sexual maturity when 8–9 months old. A wounded or frightened hare gives a high-pitched scream, but normally they are very silent animals. The familiar tracks of a hare (40d–3) are particularly easy to see in newly fallen snow. Many leverets are killed by foxes and stoats, as well as by crows and large birds of prey; a number are also killed by domestic cats.

The adult hare is regarded as a sporting animal. Hunted by packs of beagles and coursed by greyhounds, many are also shot in organized annual shoots on farms. Local populations of hares are subject to natural cyclical fluctuations over a period of approximately 10 years.

41 Mountain hare
Lepus timidus

Identification: Body-length 46–61 cm (18–24 in.); tail relatively short, 4–8 cm ($1\frac{1}{2}$–3 in.), white all over and not black on the upperside. More compact than the Brown hare and has relatively short ears with distinctive black tips; when laid forwards these reach ap-

proximately to the tip of the muzzle. Weight 2–5·8 kg ($4\frac{1}{2}$–11 oz). In most areas the winter pelage differs from that of the summer: white or greyish in winter (41a), only the ear tips remaining black, while in summer the coat is brownish (41b). Also known as the Arctic, Blue or Variable hare.

Distribution: Iceland, Faeroes (introduced 1820), Ireland, Scotland, the whole of Scandinavia except the extreme south of Sweden, and also in Finland and eastwards through the Soviet Union and Siberia to the Pacific. In addition there is an isolated population in the Alps. A closely related species occurs in northern America. There are considerable differences in size and the degree of colour change in the various, separate European populations. In Scotland and Ireland the Mountain hares are smaller than those in the Alps and Scandinavia, and the white winter coat is not so well developed. Few hares turn white in Ireland and in southern Scandinavia the winter fur is almost blue-grey.

Habitat: Found both in open country such as upland heath, moorland and mountain slopes, and also in open woodland and scrub that is predominantly deciduous, more rarely in conifer forests. In some areas the hares move to lower altitudes in winter, living in woodland rather than out in the open.

Habits: Broadly similar to those of the Brown hare (40). Like the latter they are mainly active at night, but are also often seen feeding during the day, particularly in hard weather. This is a more sociable and trusting animal than the preceding species, does not run so fast and does not make such sudden changes of direction. Its form is usually between large rocks, under a bush or in long herbage, and short runs or burrows are also dug in snow or in peat. Various green plants are eaten in summer, but the winter diet consists mainly of the shoots and bark of woody plants, such as heather, aspen, birch and willow. The Mountain hare also eats lichen in the Arctic and scrapes the snow with its fore-feet to reach food. The toes can be spread out apart from each other, thus allowing easier movement on snow. The female produces 1–3 litters of young during the course of a season, but the number of young per litter varies considerably—from 1 to 8. The gestation period is about 50 days and the young are very well developed at birth. The Mountain hare can interbreed with the Brown hare but the offspring are not fertile.

Rodents

The rodents (order Rodentia) have more species than any other order of mammals. About 3,000 species have been described so far and these are distributed throughout the world. No other mammal group shows such a wide range of adaptations for colonizing almost every type of habitat; correlated with this, their appearance and size are very varied and some of the forms show extreme specialization. All these different forms, however, have the characteristic rodent dentition. This is very similar to that of the hares and rabbits but is even more reduced. Thus, there is only one incisor in each half of the jaw; the maximum number of cheek teeth in each half of the upper and lower jaws is 5 and 4 respectively, and in the majority of forms it is 3 and 3. In many rodents the incisors are particularly large; those in the lower jaw may be so long that the part hidden in the jaw may extend backwards beyond the hindmost cheek tooth. The incisors are described as 'rootless'; they grow throughout life from an open pulp cavity and the growth rate may be as fast as several millimetres per week. The continually increasing length of the incisors is compensated for by the fast rate at which they wear: enamel is present only on the front surface, thus the back surfaces wear away more rapidly, resulting in chisel-shaped incisors which have a sharp, straight cutting edge. The enamel is often reddish-yellow to reddish-brown. The canine teeth are completely absent and there is a long, toothless gap between the incisors and the cheek teeth. In many species this gap is filled by a fold of skin from the upper lip; the mouth cavity can thus be closed without interfering with the free working of the incisors for gnawing or digging. The cheek teeth vary considerably in form: low-crowned and 'knobbly' in species which feed on a mixed diet, high-crowned or 'rootless' in those which live on tough, hard vegetation; in the latter the tooth surfaces have prominent, transverse ridges of enamel. Correlated with this the lower jaw moves backwards and forwards during chewing.

The upper lip is cleft in many species. In some the eyes and ears are very well developed, while in others they are much reduced; the eyes may even be reduced to such an extent that there is no eye-opening at all. The hind-feet have 5 digits, the fore-feet often only 4, the thumb being more or less rudimentary. The covering of fur varies from an extremely silky-soft pelt, with a thick underlayer of woolly hair, to an almost naked skin or an armour of long, powerful spines. Most rodents are essentially vegetarian and they have a well-developed caecum which functions as in the lagomorphs: a comparable 'refection' has also been observed in many rodents. In general, rodents are short-lived but they are very prolific. In many forms local populations undergo marked, cyclical fluctuations in numbers. Some species collect winter stores, while others hibernate.

In Europe the rodents are represented by 9 families or subfamilies: the squirrels (Sciuridae), the beavers (Castoridae), the dormice (Muscardinidae), the hamsters, voles and lemmings (Cricetinae), the mole-rats (Spalacidae), the true mice (Murinae), the birch mice (Dipodidae), the porcupines (Hystricidae) and the coypus (Echimyidae).

Squirrels

The family Sciuridae contains species which live in trees and others which live on and in the ground. They all have 5 cheek teeth in each half of the upper jaw—the front one may be extremely small—and 4 in the lower jaw. The tail is covered with dense fur and never has scales.

42 Russian flying squirrel
Pteromys volans

Identification: Body-length 15–17 cm (6–6¾ in.); tail 9·5–13 cm (3¾–5⅛ in.); weight 135–200 g (4¾–7 oz). Both body and tail are more flattened than in the Red squirrel (44). There is a broad fold of skin between the fore- and hind-limbs, which serves as a gliding membrane. Head small and rounded; eyes large and black; ears small and without long hairs at the tips. Fur dense, soft and with a sheen, colour greyish both in summer and winter.

Distribution: An eastern species with its western limit formed by the Baltic Sea and a line running from the northern end of the Gulf of Bothnia to the White Sea. The range extends eastwards from Finland and the Baltic countries through the Soviet Union to eastern Siberia, China and Japan.

Habitat: Mixed woodland, especially conifer forest mixed with aspen, birch and alder. Shows a preference for unmanaged forests with hollow trees but may also occur in parks and large gardens with old trees, providing there is little disturbance.

Habits: Lives by day in its nest which may be in a natural hole in an old tree, in a former squirrel's drey or wood-

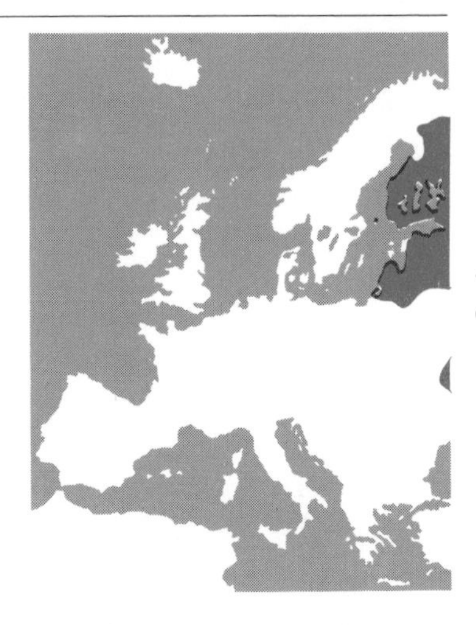

pecker's nest-hole or even in a nest-box. It does not emerge until ½–1 hour after sunset. Characterized by its ability to glide when jumping. The glide is slow, silent and follows a straight course. It will glide from the top of a tall tree and end up on the trunk of a nearby tree; the distance covered may be up to 40 m (130 ft). The food consists of the seeds of birch and alder, and also of buds, leaves, berries and fungi. Flying squirrels collect winter stores but do not hibernate. The young are born in the nest which is lined with moss, lichen, grass, rootlets and sometimes feathers. There may be 1 or 2 litters per year, each with 3–6 young. The Flying squirrel has a shrill call, which is not unlike the sound made by a Noctule bat.

43 Grey squirrel
Sciurus carolinensis

Identification: Larger than the Red squirrel (44). Body-length 24–30 cm

($9\frac{1}{2}$–12 in.); tail 10–25 cm ($7\frac{3}{4}$–$9\frac{1}{2}$ in.); weight 510–570 g (18–20 oz). Hairy ear tufts are absent during the summer and although visible in the winter, they are less pronounced than in the Red squirrel. Winter coat appears mainly grey with a yellowish-brown streak along the back; in summer it looks more brownish with rufous patches on the flanks and sometimes on the outer part of the limbs.

Distribution: Native to North America, but introduced several times to the British Isles during the period 1876–1929. It has since spread and has partly replaced the Red squirrel. The Grey squirrel is now found over large areas of England and Wales, south-west Scotland and northern Ireland.

Habitat: Deciduous and mixed woodland with oak and beech; also in conifer forests if there are seed-bearing oaks and beeches in the vicinity. Now thrives in some built-up areas where there are parks and large gardens with trees.

Habits: Much the same as those of the Red squirrel. Like the latter, it is diurnal but less shy and spends more time on the ground where it is often comparatively bold. In winter Grey squirrels can often be seen moving about in small groups and mass migrations have been recorded in North America.

In Britain the winter and breeding drey is usually built in a cleft of branches close to the tree-trunk; less often in a hollow tree. Summer dreys, more like leafy platforms, are built farther from the trunk among the outer part of the canopy. If the weather and feeding conditions are favourable this species may breed twice in the year; mating takes place in December–January and May–June. For general breeding habits, see the Red squirrel (44). Diet consists primarily of acorns and beechmast, but they also eat many other seeds and fruits as well as roots, bulbs, shoots and buds; insects and the eggs and young of birds are also taken. The Grey squirrel does considerable damage to trees by stripping the outer bark off and gnawing the inner layer of cambium and in Britain they are subject to pest control.

44 Red squirrel
Sciurus vulgaris

Identification: Body-length 19–28 cm ($7\frac{1}{2}$–11 in.); tail 14–24 cm ($5\frac{1}{2}$–$9\frac{1}{2}$ in.); weight 260–435 g (9–15 oz). Ear tips with long, hairy tufts particularly noticeable in winter. The colour of the coat and tail varies considerably, from yellowish through various shades of red and brown to almost black. A number of forms have been described. The northern forms, which include the British (44a), are smaller than the southern forms. Melanism does not occur in the British form whereas on the Continent the proportion of black squirrels is high in certain localities. The British form will interbreed with the continental form and as a result of introductions to Britain, some black specimens have been observed in the past. As well as its smaller size, the British form is characterized by changes in the pigment of the hair on the tips of the ears and tail; these changes in colour take place as the season progresses and by the summer the British squirrels have a bleached look at their extremities. Some of the continental forms develop very long ear-tufts.

Distribution: Throughout Europe where there are wooded areas. Absent, however, from certain parts of the British Isles and from the islands in the Mediterranean. The range extends eastwards through northern Asia to Japan.

Habitat: Mainly in conifer or mixed forest, but also in deciduous woodland, parks and large gardens. In Britain its preference for secluded conifer forests of mixed species may have enabled local populations to avoid being displaced by the more successful Grey squirrel, a species that shows a preference for deciduous woodland.

Habits: Strictly diurnal, lives mostly in trees and activity on the ground is usually of short duration only. It climbs rapidly up and down trees, always head first and with the tail stretched out straight behind. Moves out on to the thinnest branches and makes long leaps of 3–4 m (10–13 ft) from tree to tree. Particularly active during the morning, with another peak at dusk, but has several periods of activity during the day with intervals for resting.

In Britain the majority of dreys are built close to the tree-trunk, 30 ft up. The drey is usually smaller and more compact than that of the Grey squirrel and it has no obvious entrance. It is made of a mass of interlaced twigs and branches and lined with grass, moss and leaves. On the Continent the dreys have one or two entrances; they are usually built high up in trees but have also been found in hollow trees and deserted outhouses. Red squirrels do not hibernate, but during periods of hard weather several consecutive days may be spent in the drey.

There are usually two breeding periods, the first between February and April—sometimes as early as January—and the second in summer. The gestation period is about 6 weeks and the number of young per litter varies from 1 to 6, but is normally 3–4. The young are born naked, toothless and blind and are looked after by the female until they are about 8 weeks old. The young first breed in the following year.

Red squirrels feed mainly on the seeds of conifers, particularly Norway spruce and Scots pine. Cones stripped by squirrels (44f–g) are easily recognized: the scales are gnawed from the base upwards, giving the cone a frayed appearance, stopping just short of the tip; the top scales bear no seeds and the squirrels leave these scales intact. Acorns, beechmast, hazelnuts and other seeds are also eaten. When handling nuts a hole is first gnawed at the tip of the nut, the lower incisors are then inserted in this hole and the shell

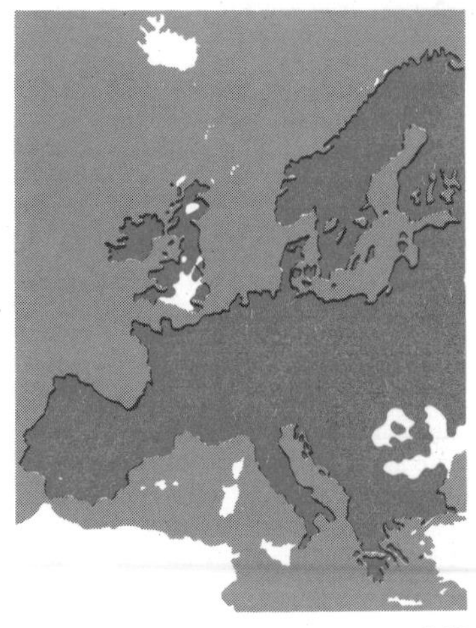

split longitudinally (44h). Red squirrels sometimes eat large amounts of tree buds and young shoots, particularly of pine and spruce, and may thus do considerable damage. Like the Grey squirrel they also damage the bark of trees. Their diet includes berries, galls and fungi as well as the eggs and young of birds. In the eastern part of their range, in the Soviet Union and Siberia, Red squirrels may undertake considerable migrations, and in this they prove to be good swimmers. They make various sounds and a scolding or chattering call is frequently heard when they are disturbed. Their major predator on the Continent is probably the Pine marten which is able to pursue them in the trees but the majority are taken at night in the nest. Large birds of prey such as the Golden eagle and the Goshawk also hunt squirrels; stoats and wild cats are among other predators but in general the Red squirrel has few enemies.

45 European souslik
Citellus citellus

Identification: Body-length 19–22 cm ($7\frac{1}{2}$–$8\frac{3}{4}$ in.); tail relatively long, 5·5–7·5 cm (2–3 in.) with very long fur; weight 240–340 g ($8\frac{1}{2}$–12 oz). Back brownish-grey, usually without distinct markings. Ears short. Also known as the Common ground squirrel.

Distribution: South-east Europe (Bulgaria, Rumania, Hungary), the adjacent regions of the Soviet Union, extending north and west to parts of Austria, Czechoslovakia and Poland. There are closely related forms in Turkey, Israel, the Caucasus, Mongolia and China.

Habitat: Originally associated with natural, grassy steppe areas, but now that these have mostly come under the plough, this species has become adapted to living high up on mountain slopes, as well as in the lowlands where it lives on banks along rivers, road and railway embankments and similar places.

Habits: Sociable, living in colonies, each individual having its own burrow in the ground. The entrances to the burrows are in soft earth, often on small hillocks. Shafts may go down vertically or obliquely to a depth of a yard or more. Towards the end of the burrow the tunnellings widen into a nest-chamber. Some of the older tunnel systems may become very complex. The Ground squirrel is essentially a diurnal animal which is active from sunrise to midday and again for a few hours in the late afternoon. It likes warm, dry weather, and will remain in the burrow while the weather is wet or overcast. Food is stored underground and before going into hibernation the entrance-hole is closed. Hibernation lasts from early autumn until March and no additional food is collected during this period.

The diet consists mainly of cereals, both the shoots and seeds being taken, but a certain amount of animal food is also eaten. Considerable damage may be done to cereal crops, both in spring and autumn. During the latter period the animals often actually live in the cornfields, digging a few temporary burrows in which to shelter.

The female produces only one litter in the year, in April or May. The number of young varies from 3 to 8, but is usually 4–6. The young are naked and blind at birth; they emerge from the burrows after 4–5 weeks but continue to be suckled by the mother for some time. The Ground squirrel is an extremely wary animal which only emerges from the burrow after making sure that everything is quiet in the open. It often sits up on its hind-legs and takes a good look all round. Various sounds are made, one of which is a high-pitched, piping whistle—a kind of alarm signal.

46 Spotted souslik
Citellus suslicus

Identification: Body-length 18·5–26 cm (7–10 in.); tail relatively short, 3·2–4 cm ($1\frac{1}{4}$–$1\frac{1}{2}$ in.) with very short fur; weight about the same as in the preceding species. Colour of back variable but there is always a distinct pattern of pale spots. Ears short. Also known as the Spotted ground squirrel.

Distribution: Found in a belt across the southern part of the Soviet Union from the Rumanian frontier in the west to the Volga in the east. To the north-west the range extends into south-eastern Poland.

Habitat and habits: Associated with grassy steppes. Habits much the same as in the preceding species.

47 Alpine marmot
Marmota marmota

Identification: A large, stoutly built rodent with a broad head and short limbs. Body-length 50·5–57·5 cm (20–22 in.); tail very long, 13–16 cm (5–6 in.); weight 4–8 kg ($8\frac{3}{4}$–$17\frac{1}{2}$ lb). Fur long and dense, variable in colour but usually dark brown, red-brown or yellow-brown.

Distribution: The Alps and Carpathians in Europe, extending eastwards into Asia and North America.

Habitat: Grassy areas from the tree limit up to the snow-limit. Prefers a rather flat terrain, but may also live on slopes provided there is an open view on all sides.

Habits: Very similar to those of the preceding species (45). Like the latter, strictly diurnal and sociable, living in colonies of varying size. The subterranean tunnels are of much larger dimensions. From the summer den, which is not particularly spacious, a number of tunnels radiate: these are approximately 1–4 m (3–12 ft) in length, 20–25 cm (8–9 in.) in diameter, and there are many side-branches and entrances. The winter den is larger, lies deeper in the earth and has only a few entrances. It measures 75 × 100 cm (30 × 40 in.) and is lined with dry grass: a smaller chamber at the end of

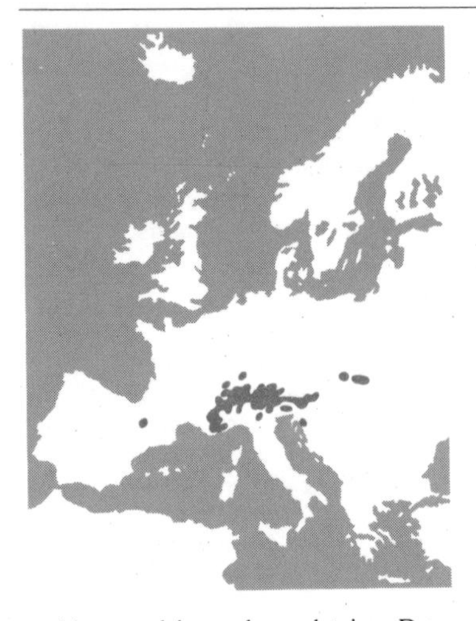

a side tunnel is used as a latrine. Dry grass is taken in and the entrance-holes are closed with earth and stones before the marmots go into hibernation in the autumn; this may last for half a year or more and during the whole of this period the animals take no food. On the other hand, during the spring and summer months they spend most of the time feeding on all kinds of fresh vegetation. Alpine marmots are very alert when they are outside and often rise up on their hind-limbs to have a look around.

Hearing and vision are well developed. They straddle their legs as they move, with the result that the gait is somewhat rocking. The females produce 2–4 young in early summer, and these remain with their parents until the next spring. There is only one litter and sexual maturity is probably not achieved until they are 3 years old. The alarm call is a high-pitched, sharp whistle. A number of other sounds are also made.

Steppe marmot
Marmota bobak

Identification: Very similar to the preceding species both in size and colour, but with shorter tail and legs.

Distribution and habitat: Formerly common on the steppes of southern Russia and north-westwards to south-eastern Poland, but the range is now much reduced.

Habits: In general similar to those of the preceding species.

Beavers

48 European beaver
Castor fiber

Identification: Europe's largest rodent. Easily recognized by the broad, flattened tail which is covered with scales. Size somewhat variable, those from Scandinavia being smaller than those from France and Germany. Body-length 74–97 cm (29–38 in.); tail 28–38 cm (11–15 in.); weight up to 35 kg (77 lb) or more. Ears very small and almost hidden in the fur, which is long and thick, varying in colour from pale to dark brown; the belly is usually paler than the back. The toes of the hind-feet are webbed and the first 2 claws are double, forming the so-called 'combing claws', which are used to keep the fur clean.

Distribution: Associated with extensive areas of forest in northern Europe and north America. Formerly much more widespread in Europe—including Britain—but it was practically exterminated in western and central Europe. As a result of protection and reintroductions populations now survive in the lower reaches of the Rhône, in the middle stretch of the Elbe and in parts of Scandinavia and

Finland. In Norway populations now thrive in the southern part of the country. In Sweden the last Beaver was killed in 1871 but after introductions in the early twentieth century (1922) they became re-established and now number several thousand. Specimens of both the Norwegian and Canadian forms have been introduced into Finland.

Habitat: Lakes and rivers surrounded by deciduous forest and scrub with trees such as ash, aspen, birch and willow, with plenty of undergrowth.

Habits: Mainly nocturnal but in secluded areas they also emerge during daylight. Adapted for aquatic life, they swim and dive well, often remaining submerged for up to 5 minutes. When active, most of their time is spent in the water. On land they move slowly and clumsily, with the tail slapping against the ground. When possible they dig burrows into the banks which have immediate access to water; elsewhere they build more elaborate structures, known as lodges, a process which often entails the construction of dams and waterways. The lodges are built either on the banks or with their foundations in shallow water. These free-standing lodges may achieve considerable dimensions: up to 2 m (6 ft) high and 4 m (12 ft) in diameter. The entrances to the lodge are below the water surface. Inside the lodge there is a large chamber which is plastered with mud and acquires a polished surface on the inside. Fluctuations in the water level are counteracted by the building of dams; the latter may be several yards long and consist of branches, stones, earth, roots and similar material, all closely packed together. Beavers also dig special channels link-

ing their lodges with feeding-places, along which they float timber from where it has been felled. They feed primarily on the bark, shoots and leaves of aspen, birch and willow. The trees are felled by the Beavers; normally they tackle stems with a diameter of 9–12 in. The work may be done by a single animal or by a pair working together. The tree is gnawed and felled with the powerful incisors (48a), which are capable of stripping off chips of wood 5–10 cm (2–4 in.) long and 3 cm ($1\frac{1}{4}$ in.) across. In the spring a wide variety of herbaceous plants is also eaten. Branches and stems of trees are collected before the winter and these stores may occupy several cubic yards. More time is spent in the lodge or burrow during the winter months but some activity still continues under the ice.

Beavers have scent glands in the genital region and territory is marked with scent. They are monogamous and a pair may live together for many years. Mating usually takes place in February and the young are born after a gestation period of about 3 months. The litter size may be 1–6 but is usually 3–4. Sexual maturity is not achieved until 3 years of age and the young remain within the family group during this time. Thus there may be more than one generation of young living with their parents even though there is only one litter in the year.

Dormice

The dormice (family Muscardinidae) are somewhat like miniature squirrels for the tail usually has long hair and the eyes and ears are relatively large. Most species live in bushes and trees and are nocturnal. They all hibernate for some months, but this is broken from time to time when they feed on stored food. Dormice have 4 cheek teeth in each half-jaw.

49 Hazel dormouse
Muscardinus avellanarius

Identification: The smallest of the European dormice. Body-length 6–9 cm (2–3½ in.); tail relatively long, 5·7–7·5 cm (2–3 in.), and somewhat bushy; weight 15–25 g (½–1 oz). Ears relatively small. Back ochre-yellow to red-brown, belly a little paler.

Distribution: Throughout most of Europe. In Scandinavia only found in southern and central Sweden. In Britain only in the south, while in Holland, Germany and Denmark the species is absent from the areas bordering on the North Sea. Not found in the Iberian Peninsula nor in certain parts of the Mediterranean region.

Habitat: Bushes, coppice and dense undergrowth.

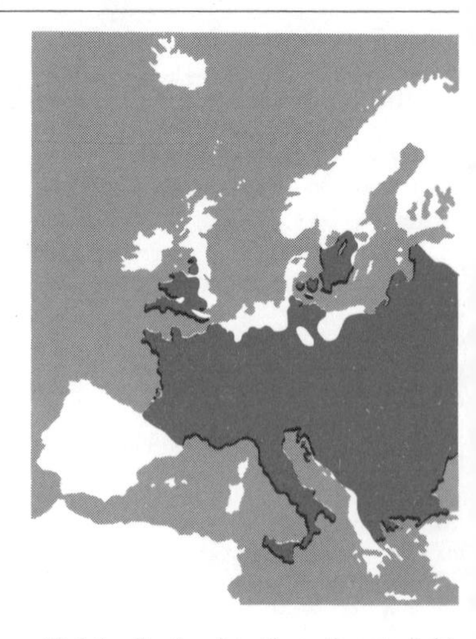

Habits: Predominantly active at night. Climbs up and down branches and twigs with great agility, turning the feet sideways and thus getting an effective grip; also leaps from branch to branch. This species is not particularly shy. Elaborate summer nests are built in trees and bushes at a height of 1–4 m (3–12 ft) from the ground. The nests are spherical or slightly oval, 10–15 cm (4–5½ in.) in diameter, with an entrance about 2 cm (¾ in.) across. They are made out of the leaves of beech, oak, hazel or other trees, which are skilfully woven together with an inner lining of bark fibre—often honeysuckle—and grass, moss, rootlets, wool and hair. The Hazel dormouse spends the day resting in the nest. There is usually only one individual in each nest, but there are often several nests within a small area. Diet is mainly vegetarian, chiefly hazelnuts, acorns, beechmast, berries, buds and shoots. Winter food is not stored but having fattened on the autumn crop of seeds and fruits these dormice are ready for hibernation. They build a winter nest in a hole in the ground, between tree roots or in a similar place. In October they roll themselves up into a ball inside the nest and usually sleep undisturbed until April–May.

A litter of young is born in June–July after a gestation period of 22–24 days; there is sometimes a second litter in August–September. The number of young per litter varies from 2 to 7 but is usually 3–4. The young emerge from the nest when they are about 30 days old and after another 10 days they become fully independent. They do not breed until the following summer. The average life span is 3–4 years.

50 Forest dormouse
Dryomys nitedula

Identification: Body-length 8–13 cm (3–5 in.); tail relatively long, 8–9·5 cm (3–3¾ in.), and bushy. Ears comparatively small. Boldly marked facial pattern with black area around each eye. Back varies from grey-brown or yellowish-grey to pale ochre-yellow.

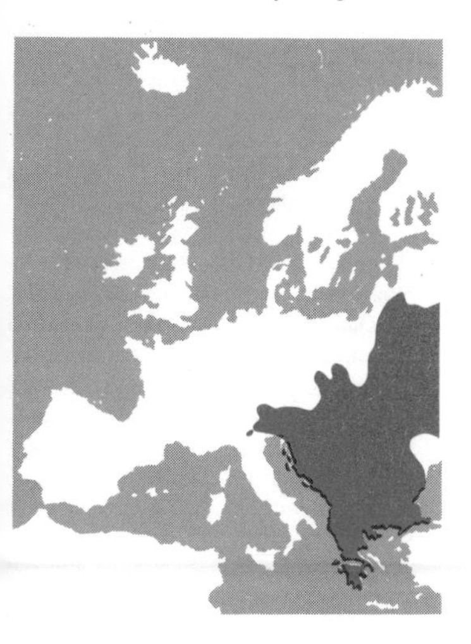

Distribution: South-east Europe, north-west to the Tyrol and Silesia. Also in the Caucasus region and far into Asia. Absent from Britain.

Habitat and habits: Deciduous and mixed forest, preferably with oak and dense bushy undergrowth. Habits approximately the same as in the preceding species.

51 Garden dormouse
Eliomys quercinus

Identification: Body-length 10–17 cm (4–6¾ in.); tail relatively long and slender, 9–12·5 cm (3½–4¾ in.), ending in a tuft of long hair; weight 45–120 g (1½–4¼ oz). Back darker than in the preceding species, ears larger and more pointed. Clearly defined facial pattern with black area around the eyes and ears.

Distribution: Portugal, Spain, France, Italy, Germany (except the areas bordering the North and Baltic Seas) and eastwards through eastern Europe to the Urals. Also in Corsica, Sardinia and Sicily. Related forms in the Balearic Islands. Absent from Britain, Denmark and Scandinavia.

Habitat: Deciduous and mixed woodland. Also in orchards, parks and large gardens. Often enters cellars and store-rooms, especially in the autumn. Also associated with rocky terrain.

Habits: Mainly nocturnal, like the other dormice. Although an agile climber this species spends a lot of time on

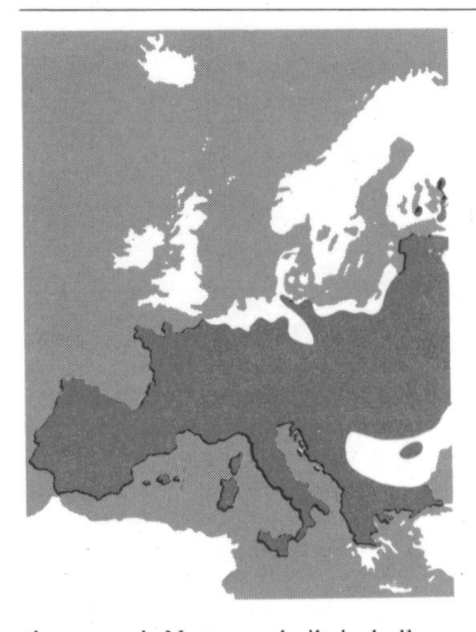

the ground. Nests are built in hollow trees, nest-boxes, rock crevices and sometimes even an old bird's nest in a bush is used as a foundation. Normally it hibernates from October–November to April–May, hidden in a hollow tree, wall cavity, under a rock or in a similarly sheltered spot. When living in buildings hibernation does not take place. The gestation period is 22–28 days. There may be 1–2 litters per year, the size varying between 2 and 7 young. The average life-span is $5\frac{1}{2}$ years. Diet consists of animal and plant food, including fruit and seeds. Insects and snails play an important role, and observations have shown that they are capable of killing birds and small mammals. It is also possible that food is stored for the winter.

52 Edible dormouse
Glis glis

Identification: The largest of the European dormice. Body-length 13–19 cm ($5–7\frac{1}{2}$ in.); tail long, 11–15 cm ($4\frac{1}{4}–6$ in.), and very bushy, the same colour as the body; weight 70–180 g ($2\frac{1}{2}–6\frac{1}{4}$ oz). Back grey to brownish-grey. No clearly defined facial pattern although there is a dark ring around each eye.

Distribution: Approximately the same as that of the preceding species. In Spain, however, it is confined to the north and in France it is found only in the south. Introduced into England in 1902 and it now occurs in a limited area to the north of London, along the Chiltern Hills. Not recorded in Denmark, Norway, Sweden or Finland. The range extends eastwards to Iran and Turkestan.

Habitat: Mature wood, preferably deciduous, orchards, parks and large gardens. Particularly numerous where there are plenty of hollow trees and good fruit crops. Often enters houses in autumn.

Habits: Mainly nocturnal but is also seen during the day. A very fast, active animal which mostly lives in trees and bushes. Climbs with agility and jumps 7–10 m (23–32 ft). Moves right up into the leaf canopy of tall trees and out on to the ends of the branches, to get at seeds such as beechmast. On still summer nights their presence is often betrayed by continual rustling and rasping sounds heard overhead and several individuals are active within a small area. Fights often take place, especially among males at mating time. During the summer they usually spend the day lying up in a nest some height above the ground. These nests are not elaborate and are always in cavities, such as a hollow tree, between stones in a wall or in a nest-box. Winter nests are usually burrowed into the ground under tree roots or into the rotting wood of a hollow tree. Buildings also provide suitable sites, usually in cellars or roof cavities. Several individuals will often sleep close together. They remain in hibernation from October to April. During this period, having fattened in the autumn, they lose up to half their weight and the body temperature drops to about 4° C (40° F). In the eastern part of the range hibernation lasts for at least 8 months. Diet includes a wide variety of fruits, seeds and berries as well as insects and other small animals. Considerable damage may be caused in orchards and forestry plantations, particularly among young conifers.

Mating takes place in May–June in central Europe. The young are born about a month later. There are frequently 4–5 young although the number varies from 1 to 8. Sexual maturity is not reached until the second year and there appears to be only one litter per year.

Mouse-tailed dormouse
Myomimus personatus

Identification: A little smaller than the Forest dormouse (50) and without the bold pattern of black around the eyes; also the tail is not bushy but tapering. A grey dormouse with short white hair on the tail.

Distribution: First described in 1924 from south-western Turkmenistan near the frontier with Iran, and later also discovered in south-eastern Bulgaria.

Habitat and habits: Apparently not very different from those of the other dormice but it may be exclusively terrestrial.

Hamsters, lemmings and voles

The numerous members of the subfamily Cricetinae differ from the true mice (Murinae) in having a shorter muzzle, a broader head, smaller eyes and ears, and shorter legs and tail; the latter is covered with hairs. Each half-jaw has 3 cheek teeth with persistent pulp cavities. The chewing surfaces of the teeth have alternating folds of enamel and dentine, forming a zigzag pattern of ridges. Most species are terrestrial and the diet consists of coarse plant food.

53 Common hamster
Cricetus cricetus

Identification: A medium-sized, clumsily built rodent. Body-length 21·5–32 cm ($8\frac{1}{2}$–$12\frac{1}{2}$ in.); tail short, 2·8–6 cm (1–$2\frac{1}{4}$ in.), with short hairs; weight

150–385 g ($5\frac{1}{4}$–$13\frac{1}{2}$ oz). Unique among European mammals in being darker below than above.

Distribution: Mainly in central and eastern Europe—including the Caucasus—extending eastwards to Lake Baikal and westwards to Belgium.

Habitat: Originally associated with grassy steppe country, but has become adapted to living in cultivated areas.

Habits: Mainly active above ground at night, rarely during daylight. Hamster populations are usually dense in the areas where they settle, but each individual is solitary, and has its own burrow in the earth. Sometimes the burrow goes down to a depth of about 2 m (6 ft). There is a system of tunnels and chambers. The tunnels have a diameter of 6–8 cm (2–3 in.); some of them go down vertically, others are only slightly sloping. Deep down there are a number of chambers; one of these, lined with soft material, is the true nest. The others serve as a latrine and as storage chambers. Hamsters have large cheek pouches and in late summer and autumn they carry food in these, transporting large quantities of beans, peas, corn and other seeds into the storage chambers. Before entering hibernation they close the entrance-holes with earth. Hibernation is broken at frequent intervals when the animal eats some of its stores.

The breeding season extends from spring to autumn. The number of litters is not known, but there are probably 2 per year; the number of young is usually 6–12 but litter-size varies from 4 to 18. The gestation period is about 20 days. Hamsters live for about 2 years. In spite of their apparent clumsiness, they move extremely fast; they run, jump and climb remarkably well, and are expert diggers. They are very aggressive towards members of their own species and towards other animals. Scent secreted from glands on the flanks probably plays some part in marking individual territories and there is also a scent gland in the region of the abdomen.

Golden hamster
Mesocricetus auratus

Identification: Somewhat smaller than the Common hamster (53). Body-length 15–18 cm (6–7 in.); tail very short, 1·2–2 cm ($\frac{1}{2}$–$\frac{3}{4}$ in.). Similar in colour to the preceding species, but not black on the belly.

Distribution and habitat: The Golden hamster comes originally from Syria, but it is closely related to *M. newtoni* which occurs in Bulgaria, Rumania and the Ukraine, where it is associated with uncultivated steppe country.

Habits: Commonly kept as pets and laboratory animals. Particularly suitable for laboratory work as the gestation period is as short as 16 days, and the female can have 5 litters of 8 young in a year. In addition they can start to breed at an age of 2–3 months. Little is known about their habits in the wild.

Grey hamster
Cricetulus migratorius

Identification: Much smaller than the Common hamster (53). Body-length 8·7–11·7 cm ($3\frac{1}{4}$–$4\frac{1}{2}$ in.); tail 2·2–2·8 cm (about 1 in.). Back grey-brown, belly white.

Distribution: Soviet Union and western Asia, with an apparently isolated population in Greece.

Habitat and habits: Originally associated with tree-covered slopes and scrub in steppe country, but may also be found in fields and gardens. Habits, in general, as for the Common hamster.

54 Wood lemming
Myopus schisticolor

Identification: Body-length 8·5–9·5 cm ($3\frac{1}{4}$–$3\frac{1}{2}$ in.) (some sources give up to 12·5 cm or $4\frac{3}{4}$ in.); tail short, only 1·5–1·9 cm ($\frac{1}{2}$–$\frac{3}{4}$ in.); weight 10–45 g ($\frac{1}{4}$–$1\frac{1}{2}$ oz). Almost uniformly slate-grey, with slightly paler underparts and a red-brown area on the rear of the back.

Distribution: Conifer forest areas from Scandinavia and Finland through the northern parts of the Soviet Union and Siberia to the Pacific Ocean. In Scandinavia the range extends southwards to Oslo Fjord; in Finland it is absent from coastal areas in the south and west.

Habitat: Conifer forest with thick moss cover, preferably with secondary growth and fallen branches.

Habits: Unobtrusive and retiring, remains hidden during daylight. Main activity takes place after dark. Burrows and runs are made in the moss cover. A spherical nest of moss and grass is built either among the roots of a tree or hidden under fallen branches. The first litter is born at the beginning of June. There may be 3–7 young. A second litter (sometimes a third) is born later in the summer. Wood lemmings feed chiefly on mosses but also eat lichens and liverworts. Their presence is often betrayed by bare patches in the moss cover. They do not hibernate. In certain years they occur in large numbers and undertake mass migrations, but these are not so extensive as in the following species.

55 Norway lemming
Lemmus lemmus

Identification: Body-length 13–15 cm (5–5¾ in.); tail very short, 1·5–1·9 cm (½–¾ in.); weight up to 110 g (3¾ oz). Conspicuously mottled with shades of rust-brown, fawn and varying amounts of black.

Distribution and habitat: Mountain regions in Scandinavia from Hardanger in the south to Finmark in the north, and eastwards along the coasts of the Arctic Ocean to the Kola Peninsula. Moves down in winter into the birch and conifer forests, and in mass migration years it extends well into the lowlands. There are related forms farther east in the northern parts of the Soviet Union and Siberia.

Habits: In certain years this species occurs in very large numbers and lemmings are seen all over the mountains down to the conifer forest belt, whereas in other years it may scarcely be seen at all. This concentration of numbers occurs about every fourth year—hence the expression 'lemming year'. It is thought to be due to a combination of factors. Changes in the number of predators are significant; also, when climatic conditions are favourable and food is abundant, there is an increase in the lemming's rate of breeding, and this may even continue in winter. As numbers rise the density becomes too great and the lemmings undertake mass migrations. They leave the overcrowded areas and spread out in all directions, apparently moving at random and tackling all sorts of obstacles. They may even try to cross the sea. Nearly all these migrants die. Lemmings are not averse to carrion and the survivors have been observed feeding on those which have died.

In winter, lemmings mostly live in dry places on the mountains, where they make extensive runs beneath the snow. During the summer they move into damper areas, such as marshland with scrub of dwarf birch and willow. Here they build large round nests either among vegetation or in crevices under rocks. They feed on all kinds of vegetable matter, mosses playing an important rôle. Breeding normally starts in April–May, but as already mentioned it may extend into the winter. After a gestation period of about 18–20 days the female gives birth to a litter of 5–7 young. In a good year there may be 4–5 litters and possibly an extra one in winter. The young are independent at an age of 3 weeks and become sexually mature a few weeks later.

56 Bank vole
Clethrionomys glareolus

Identification: Body-length 8–12·3 cm (3–4¾ in.), but size varies greatly according to locality; tail bicoloured and relatively long, 3·6–7·2 cm ($1\frac{1}{4}$–$2\frac{3}{4}$ in.) or about half the body-length; weight 16–30 g ($\frac{1}{2}$–1 oz). Back reddish-brown, gradually becoming greyer on the sides, belly whitish-grey. In this and the following 2 species the enamel ridges on the teeth are rounded (56a), and without the angular ridges found in the Common vole and its relatives (61a–66a). In these 3 species (56–58), the ears are also more conspicuous than in the Field vole (62).

Distribution: Throughout Europe—although not continuous—and eastwards into western Asia. Absent from most of the Iberian Peninsula and other parts of southern Europe. In Scandinavia it reaches latitude 68° N. In the British Isles it is absent from certain islands and it has only been found in one area of Ireland, in the south-west.

Habitat: Associated with wooded country, particularly where there is plenty of undergrowth, and on more open ground providing there are bushes, hedgerows or an adequate shrub layer. Prefers deciduous woods to conifers but is also found in mixed woodland.

Habits: Most active at dusk and dawn. It is not a particularly retiring species and providing there is adequate cover it moves about during the day as well as after dark. Runs fast on the ground and climbs with agility, going to a considerable height in trees. Active use is made of a system of runs at a depth of 2–10 cm ($\frac{3}{4}$–4 in.); there are many entrance-holes and at the end of the runs there are small chambers where the voles lie up. The breeding nest, made of grass and leaves, is very large,

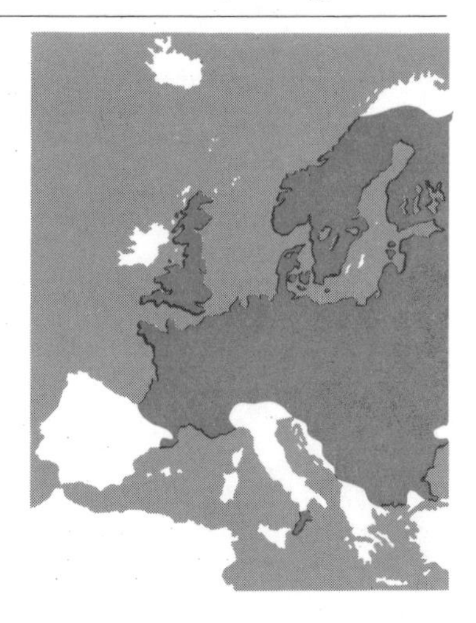

20–30 cm (8–11 in.) in diameter. Sometimes it is built on the ground, in a tree-stump or under a rock, but usually it is below ground at a depth of 40 cm. (15 in.). Bank voles forage above ground and their diet includes animal as well as plant food. Grasses and other green plants are important during the spring; insects, fruit and various seeds are also part of their diet. They may do considerable damage in forestry plantations, especially in winter, by destroying cones and gnawing bark, particularly of ash and larch. Where bark damage occurs at a considerable height from the ground this is due to the Bank vole and not to the Field vole. In the northern part of their range they store food in the winter and often enter houses and other buildings.

Breeding usually starts in April–May and lasts until September–October. It may continue even later in the southern part of their range. The gestation period is about 18 days and

4–5 litters may be produced in the same year. The size of the litter varies from 3 to 6. The young weigh about 2 g ($\frac{1}{15}$ oz) at birth, are independent after about 18 days, and may be sexually mature at an age of 4–5 weeks. The life-span is 1–2 years. Cyclical fluctuations in numbers are not as marked in this species as in the Field vole.

57 Northern red-backed vole
Clethrionomys rutilus

Identification: About the same size and colour as the preceding species. Body-length 9·8–11 cm ($3\frac{3}{4}$–4 in.), but tail considerably shorter, 2·3–3·5 cm ($\frac{3}{4}$–$1\frac{1}{4}$ in.) or about a third of the body-length. The tail is more hairy than in the following species.

Distribution: From northern Norway, the most northerly parts of Sweden and Finland eastwards through the northern Soviet Union and Siberia

to the Pacific Ocean; the North American form should probably also be regarded as belonging to the same species. In Scandinavia the southern limit is approximately about the northern limit of the Bank vole (56); both species can be found in the same localities in the area of overlap.

Habitat: In the fir and birch wood zone in places with dense undergrowth. Often enters houses in winter.

Habits: Not significantly different from those of the Bank vole (56).

58 Grey-sided vole
Clethrionomys rufocanus

Identification: A little larger than the two preceding species. Body-length 11–13 cm ($4\frac{1}{4}$–5 in.); tail relatively short, 2·8–4 cm (1–$1\frac{1}{2}$ in.), or about a third of the body-length, and less hairy than in the preceding species.

Distribution: From the mountain regions of Norway and Sweden through north Finland and the northern parts of the Soviet Union and Siberia to the Pacific Ocean.

Habitat: In the areas of lichen and scrub beyond the forest belt, varying from dry, almost bare ground to damp areas with comparatively lush vegetation. Often enters houses in winter.

Habits: This is probably the commonest small rodent in the mountains of Scandinavia. It is not a particularly retiring species. It moves about very slowly and climbs well. Runs are made under moss and lichen and in other vegetation. In winter this species lives below the snow, where it builds nests of dry grass. The summer and breeding nests are built on the ground,

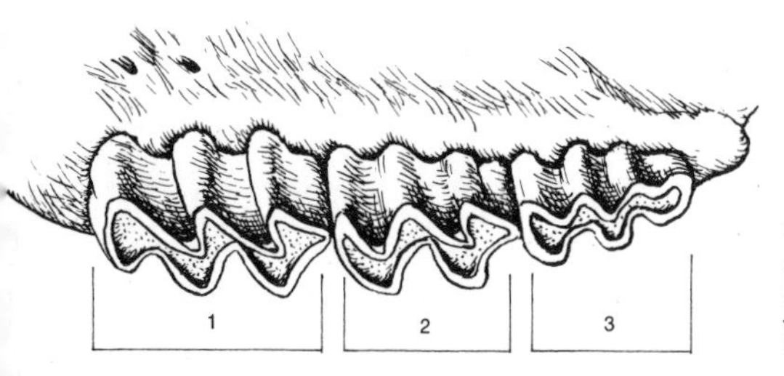

Water vole. Cheek teeth in the left upper jaw, seen diagonally from below.

under rocks, roots and in similar places. These nests are about 15 cm ($5\frac{1}{2}$ in.) in diameter. Diet consists of leaves, buds, shoots, bark, berries and seeds of various kinds, but also includes a certain amount of animal food, such as insects. In winter they climb into dwarf birch and willow and gnaw the bark and twigs. The breeding period lasts from April–May to August–September. There are usually 2–3 litters per year, each with 5–7 young. The size of the population fluctuates considerably from year to year.

59 Water vole
Arvicola terrestris

Identification: Body-length 12–22 cm ($4\frac{3}{4}$–$8\frac{3}{4}$ in.); tail 5·6–10·4 cm (2–4 in.). These measurements refer to specimens from central Europe and serve only as an indication because the size of this species varies as between one region and another. Water voles in Britain, for instance, are larger than those in central Europe but also vary in size and colour in different parts of Britain. On the basis of external characteristics it is very difficult to distinguish *Arvicola terrestris* from the next species, *A. sapidus*, although the tail in the former is relatively shorter.

Distribution: Throughout Europe except for western and southern France, Spain, Portugal and other southern areas of Europe. Extends into Asia eastwards to the River Lena and Iran. Absent from Ireland.

Habitat: In the northern parts of its range—which includes Britain—this species is associated mainly with the banks of streams, rivers, lakes and marshes, where there is dense vegetation and scattered trees, but also occurs some distance from water in fields, gardens and especially in orchards. In the southern parts of its range this species is mainly associated with grassland.

Habits: Mainly active in the morning and evening, but is also seen during the day. It runs very slowly and does not climb well, but it is a remarkably good performer in the water. Extensive tunnel systems are made along waterside banks and these are usually not far below the surface. There may be several burrows running parallel with the bank and connected by a number of cross-tunnels, many of which often end in openings near the water. Sometimes the ground is so riddled with tunnels that the surface caves in when walking over it. Small platforms, used as feeding-places, are often built out in the water. In fields and gardens where the ground is normally firmer, the burrows are oval in cross-section, the greatest diameter being in the vertical plane (in contrast to mole tunnels). Mounds of earth are thrown up at intervals and there are a number of open holes. Individual territories are marked by scent which is transferred to the hind-feet from glands on the flanks. The latter are prominent and can be seen in both sexes throughout the year.

Grass is an important item in their diet although the leaves and stems of many flowering plants are eaten, together with various kinds of seeds, bulbs and roots; the bark and other parts of woody plants are also eaten. There is some evidence to suggest that animal food is acceptable even though it does not appear to be part of their regular diet. Water voles collect winter stores, carrying them underground into spe-

cial chambers. In some parts of thei range food is stored in considerabl quantities. Damage to farming inter ests occurs in places where the vole burrow into stores of root-crops an gnaw the roots of fruit trees. Nests ar built underground or in sheltered site above ground. The latter include reed beds, under bushes, inside rotten log lying on the ground and even at th side of a swan's nest. Breeding usuall starts in April–May, and continues as in other voles, until September-October. The gestation period is abou 3 weeks. There may be 2–4 litters pe year; the litter-size varies from 4 to 8 The young weigh about 5 g ($\frac{1}{6}$ oz) a birth and reach a weight of abou 130 g ($4\frac{1}{2}$ oz) in 50 days. By this tim they are fully independent and th young of early litters may breed in th same year. In some parts of thei range, however, breeding may not tak place until the second year. The life span is short, 15–20 months at th most.

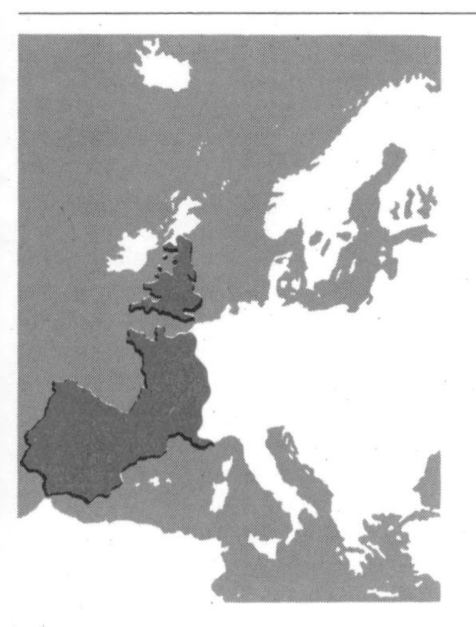

60 South-western water vole
Arvicola sapidus

Identification: Very similar to the northern form of *A. terrestris* (59). Body-length 16·2–22 cm ($6\frac{1}{2}$–$8\frac{3}{4}$ in.); tail 9·8–14·4 cm ($3\frac{3}{4}$–$5\frac{1}{2}$ in.), relatively longer than in the preceding species; weight 150–280 g ($5\frac{1}{4}$–$9\frac{3}{4}$ oz).

Distribution: Throughout the Iberian Peninsula and also in western and southern France. (Owing to the difficulty in distinguishing between *A. sapidus* and *A. terrestris* there is still some confusion about their respective ranges in France.)

Habitat and habits: Associated with water; not significantly different from those of *A. terrestris.*

Martino's snow vole
Dolomys bogdanovi

Identification: A very large, blue-grey vole. Body-length 10–15 cm (4–$5\frac{3}{4}$ in.); tail 7–12 cm ($2\frac{3}{4}$–$4\frac{3}{4}$in.).

Distribution: Restricted to a few areas in the mountains of southern Yugoslavia.

Habitat and habits: On the higher mountain slopes in the calcareous areas with grassland and scattered scrub down to an altitude of 600 m (1960 ft). Very little is known about the habits of this species which is mainly active after dark.

61 Common vole
Microtus arvalis

Identification: About the same size as the next species. Body-length 9·5–12 cm ($3\frac{3}{4}$–$4\frac{3}{4}$ in.); tail 3–4·5 cm (1–$1\frac{3}{4}$ in.); hind-foot usually not more than 17 mm ($\frac{11}{16}$ in.) long; weight 14–46 g ($\frac{1}{2}$–$1\frac{1}{2}$ oz). Hair on ears shorter and more dense than in the next species. Cheek teeth: see fig. 61a. Note that of the 3 cheek teeth in the upper jaw the middle one has only 4 enamel ridges, and that the foremost cheek tooth in the lower jaw has 7–9 enamel ridges.

Distribution: Range extends from north Spain, France and the North Sea countries in a broad belt eastwards through central and eastern Europe to parts of northern Asia. Absent from Scandinavia and most of Italy and the Balkans. This species is absent from the British Isles except for two forms which are of larger size and found only in the Orkney Islands and on Guernsey in the Channel Islands.

Habitat: In grassland—preferably where it is grazed or kept fairly short—such as meadows, fields, gardens and orchards. Thus, the Common vole, unlike the Field vole (62), is

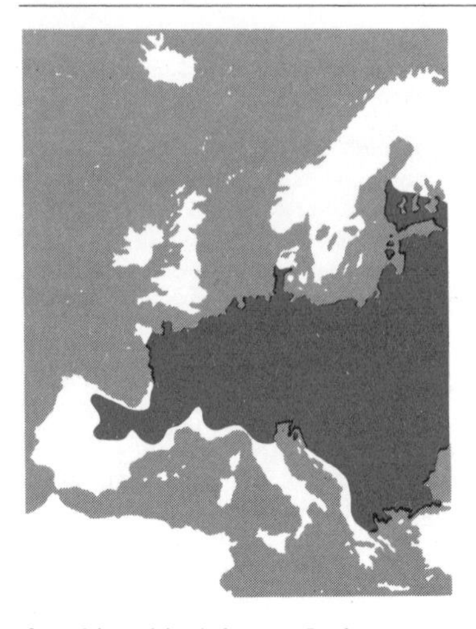

found in cultivated areas. In the mountains of central Europe it reaches an altitude of 2,300 m (7,500 ft).

Habits: Emerges both in daylight and darkness. The peak of activity is at dusk; in daylight it is more active in the early part of the day and at night there are long periods of inactivity. Extensive tunnel systems are excavated in the ground, which connect with runways in short vegetation. There are a number of chambers associated with the shallow burrows and some go down to a depth of about 50 cm (20 in.). This is a fast, lively animal which often stops in its tracks and rises up on its hind-legs. It swims readily. The senses of sight, hearing and smell appear to be well developed. Common voles usually live in very dense populations but they cannot be described as social animals. This vole appears to be very territorial and during the mating season there are frequent fights as the males go from one female to another. A nest of dry grass is usually built underground but sometimes above ground. Fertility is high: there may be as many as 10 litters per year, although 5–7 is more usual, each with 3–7 young. The latter become sexually mature when very young and may breed after a few weeks, providing the population density is not too great. The gestation period is 19–21 days. In winter a number of voles may share the same underground chamber and in severe weather stored food is eaten underground. Diet consists mainly of the green parts of plants, particularly grasses and clover, but the seeds and bark of trees and bushes are also eaten. Their subterranean habits may result in economic damage on farmland, particularly where systems of dykes and banks may be weakened or undermined. Sometimes a population will migrate to a new locality. Barns, stacks of straw and so on may be invaded in autumn and, providing the population settles in, breeding may then continue through the winter. In certain years this vole occurs in enormous numbers, reaching a peak which is then followed by a sudden mass mortality. When alarmed the Common vole utters high-pitched squeaks which are easily distinguished from the much louder and lower-pitched chattering sounds made by the Field vole.

62 Field vole
Microtus agrestis

Identification: Approximately the same size as the preceding species, but the tail is relatively a little shorter. Body-length 9·5–13·3 cm ($3\frac{3}{4}$–5 in.); tail 2·7–4·6 cm (1–$1\frac{3}{4}$ in.); hind-foot usually over 17 mm ($\frac{11}{16}$ in.) long;

weight 19–52 g ($\frac{3}{4}$–$1\frac{3}{4}$ oz). Hair on ears longer but less dense than in the preceding species. Cheek teeth: see fig. 62a. Note that of the 3 cheek teeth in the upper jaw the middle one has an extra (fifth) enamel ridge at the rear and that the front cheek tooth in the lower jaw usually has 5 angles on the outside.

Distribution: Widespread in Europe, extending eastwards to Lake Baikal and the River Yenesei and southwards to the Alps and Pyrenees, but is replaced by other species of vole in parts of southern Europe and Scandinavia. In Britain, where it is also known as the Short-tailed vole, it is widespread on the mainland but absent from certain islands. It does not occur anywhere in Ireland.

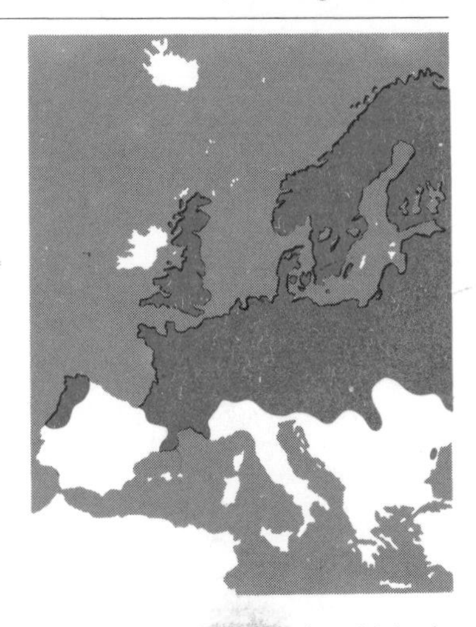

Habitat: Among tall grass, along the edges of fields and hedgerows and in young plantations of trees where the grass is lush. Prefers rather damp sites but is also found on dunes and heathland. It also occurs on high moorland where there is adequate cover from heather and scree.

Habits: Similar, in general, to those of the Common vole, but the Field vole tends to be less subterranean and is more often seen in daylight. Where populations are dense the network of runs in the vegetation is often quite conspicuous. Green droppings are deposited in piles at various places in the runs and small heaps of stems indicate where the voles have been feeding on grasses, sedges and rushes. There are also underground runs which are very close to the surface. It is likely that the droppings serve as territorial marks since it is known that adult males have well-developed scent glands in the genital region. Males in the breeding season also show aggressiveness by making loud chattering sounds. Nests may be below or above ground but in the latter case they are always well hidden, either in vegetation or in the shelter of a fallen log or similar object. The breeding habits are much the same as those of the Common vole but fertility may be slightly lower. Serious damage may be done by the Field vole, especially in winter, by gnawing at the roots of young trees or by barking saplings close to the ground. Young fruit trees and other plantations may suffer and the damage is particularly serious in years when the local population is at its highest. As in the preceding species, numbers tend to rise to about every third or fourth year and then undergo a sharp decline. Field voles have many enemies and provide an important source of food for nearly all predators, both mammalian and avian, in the area.

63 Northern vole
Microtus oeconomus

Identification: A little larger and darker brown than the Field vole (62). Body-length 11·8–14·8 cm ($4\frac{1}{2}$–$6\frac{1}{4}$ in.); tail relatively longer, 4–6·4 cm ($1\frac{1}{2}$–2 in.); hind-foot over 17 mm ($\frac{11}{16}$ in.) long; weight 24–62 g ($\frac{7}{8}$–$2\frac{1}{8}$ oz). Hair on ears comparatively long and sparse, as in the Field vole. Cheek teeth: see fig. 63a. Note that of the 3 cheek teeth in the upper jaw the middle one has only 4 enamel ridges, as in the Common vole (61), and that the foremost cheek tooth in the lower jaw has only 4 angles on the outer side. It is possible that this species should be known as *M. ratticeps*.

Distribution: From Germany and northernmost Scandinavia eastwards through the northern parts of the Soviet Union and Siberia to the Pacific Ocean; also in Alaska and western Canada. There are also scattered populations in southern Norway, central Sweden, Holland, Belgium and Hungary. Absent from the British Isles.

Habitat: Usually in damp places such as fenland, marshes, meadows and in fields liable to flooding. Also in marshy ground along river banks.

Habits: This species swims well. Like other voles it forms tunnel systems with nests and storage chambers underground and in vegetation. Where the population is dense, there are often numerous small mounds of earth, resembling molehills. The nest may be built above or below the surface of the ground, often in a grass tussock or among reeds. Diet consists of grass, roots, bark, leaves and berries. There are 2–4 litters per year, each with 5–9 young.

64 Alpine vole
Microtus nivalis

Identification: A little larger than the Field vole (62). Body-length 11·7–14 cm ($4\frac{1}{2}$–$5\frac{1}{2}$ in.); tail long, 5–7·5 cm (2–$2\frac{3}{4}$ in.), and almost completely white; hind-foot 18·5–22 mm ($\frac{3}{4}$–$\frac{7}{8}$ in.) long; weight 38–50 g (about $1\frac{1}{2}$ oz). Colour grey, coat long and dense. Cheek teeth: see fig. 64a. Note the characteristic shape of the front enamel ridges of the foremost cheek tooth in the lower jaw.

Distribution and habitat: Mainly in alpine and subalpine mountain areas in the Pyrenees, Alps, Carpathians, Apennines and various ranges in the Balkans. Lives in open rocky terrain where there are patches of alpine vegetation, also among scattered trees and bushes.

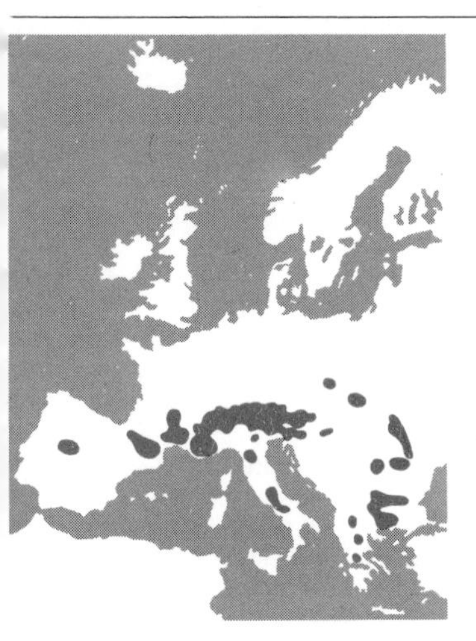

Habits. Like other voles, this species forms a system of runs, using holes and cavities under and among rocks. Often seen during the day and it is not a particularly retiring species. Swims, climbs and jumps readily. It is said to look comparatively long in the leg as it runs over the ground, carrying its tail high.

Diet consists of grasses and other alpine vegetation, including the twigs of bilberry (*Vaccinium myrtillus*). Breeding as in other voles, but the season is probably less protracted.

Mediterranean vole
Microtus guentheri

Identification: A large vole, very similar to the Common vole in external appearance, but with a short tail (about a quarter of the body-length).

Distribution and habitat: Mainly in the eastern Mediterranean, extending into North Africa.

Habits: Approximately as in the Common vole.

65 Northern root vole
Pitymys subterraneus

Identification: Body-length 7·5–10·6 cm (3–4$\frac{1}{4}$ in.); tail 2·5–3·9 cm (1–1$\frac{1}{2}$ in.); hind-feet usually under 16 mm ($\frac{5}{8}$ in.); weight 12–24 g ($\frac{1}{3}$–$\frac{7}{8}$ oz). Cheek teeth: see fig. 65a. The species of *Pitymys* differ from those of *Microtus* in having smaller eyes and ears, characters which suggest that they are more highly specialized for subterranean life. The tail is short. The hind-foot has 5 pads as against 6 in *Microtus*, and the females have fewer nipples. In *Pitymys* the foremost cheek tooth in the lower jaw is characteristic of the genus. The systematics of the European species of *Pitymys* are very difficult. Several forms have been described, but the exact differences between the species are still not clear. The colour of the fur is too variable for identification purposes. The three species illustrated here, however, can be readily identified by the shape of the rearmost cheek tooth in the upper jaw.

Distribution: Widespread but not continuous from north-west France and Belgium in a broad belt eastwards through central and eastern Europe. Absent from northern and southern Europe, and from the British Isles.

Habitat and habits: Active throughout the year. The main periods of activity are in the late afternoon and at night, although some activity also takes place during the day. They live in grassland where the soil is suitable for excavating burrows and there is plenty of vegetation. The Northern root vole is found mainly in damp meadows and fields, in banks, along canals and rivers, and also in comparatively open terrain with scattered trees. At the southern extent of its range it is found at comparatively high altitudes. All the species are good diggers and use systems of tunnels with chambers for nests and storing food. These underground systems go down to a depth of 35 cm (13 in.), although usually they are closer to the surface. The tunnels are approached by a number of entrance-holes which are well hidden by a rock or under a tussock. It is characteristic of these species that they live in colonies. Sometimes the entrance-holes are so close together that the earth thrown up by the excavations makes a dark patch in the green surroundings. Thus a colony may do considerable damage. Diet consists of the leaves, stems and roots of grasses and other plants, and large amounts of food are collected for winter storage. Precise data on breeding are not available but apparently there are several litters per year—up to 9—although the number of young per litter is probably 2–3, and therefore somewhat lower than in the species of *Microtus*.

66 Mediterranean root vole
Pitymys savii

Identification: Body-length 8·2–10·5 cm ($3\frac{1}{4}$–$4\frac{3}{16}$ in.); tail 2·1–3·4 cm ($\frac{3}{4}$–$1\frac{3}{8}$ in.). The ears are even smaller than in the preceding species and the limbs are rather longer. Cheek teeth: see fig. 66a. Note that there is a considerable difference in the shape of the last cheek tooth in the upper jaw in comparison with the preceding species.

Distribution: From north-west Spain through south France, Italy and Sicily and into the Balkans, where there are scattered populations at varying altitudes.

Habitat and habits: Little information—refer to the preceding species.

67 Iberian root vole
Pitymys duodecimcostatus

Identification: Body-length 9·3–10·7 cm ($3\frac{5}{8}$–$4\frac{1}{2}$ in.); tail 2–2·9 cm ($\frac{5}{8}$–$1\frac{1}{8}$ in.). Colour of coat very variable. The last cheek tooth in the upper jaw resembles that of the Mediterranean root vole (66).

Distribution: Portugal, Spain and southernmost France (Languedoc and Provence), but not in the Pyrenees.

Habitat and habits: Little information—refer to Northern root vole (65).

168 Musk-rat
Ondatra zibethicus

Identification: Body-length 26–40 cm ($10\frac{1}{4}$–$15\frac{3}{4}$ in.); tail 19·27·5 cm ($7\frac{1}{4}$–$9\frac{3}{4}$ in.) or about three-quarters of the body-length and very powerful; it is laterally compressed, covered with scales and only very few hairs; weight up to 1·7 kg ($3\frac{1}{2}$ lb). Ears almost hidden in the fur. Hind-feet with a narrow web between the inner parts of the toes. Back yellowish-brown, underparts greyish.

Distribution: Native to north America, whence a few pairs were introduced into Bohemia in 1905. Due to its value as a commercial fur—known as musquash—the Musk-rat was later introduced into several European countries, as well as into Japan and the U.S.S.R. It is now feral in Finland and Sweden, in many areas in central and eastern Europe, and in some parts of western Europe.

It was introduced to the British Isles about 1929 and colonies became established in several places, including Ireland, but it became a serious agricultural pest and was finally exterminated in 1937.

Habitat: Prefers ponds and small lakes where the water is comparatively calm and there is a dense growth of aquatic vegetation. Also occurs along the banks of large lakes, canals and rivers with marsh verges and low islands, where the vegetation is lush.

Habits: Slow-moving on land but thoroughly at home in the water. Swims rapidly with only the head and the front part of the back showing above water, propelled by the hind-limbs and tail while the fore-limbs are held close to the body. It also dives well and can remain submerged for several minutes. Where the banks are high enough burrows are excavated in the earth. The entrances often lie below the water surface. The tunnels lead upwards and may extend for a distance of 2–10 m (6–32 ft) from the water, depending on the slope of the ground. A nest chamber with a diameter of 25–35 cm (10–14 in.) is built in the tunnel system, just below the surface of the ground. The nest is lined with leaves, grass and moss; the site of the nest chamber is dry, at least for the duration of the breeding season. Where the banks are low and swampy, 'lodges' may be built—similar to those of the Beaver—made of twigs and the stems of grasses, reeds and rushes. Both sexes take part in the construction which may reach a height of 1 m (3 ft). Each system of tunnels or 'lodge' is occupied by a single pair of adults with young. Early litters become independent after about a month but may remain in the vicinity of the parents for a time. Musk-rats are territorial and it is possible that territory is marked with scent from glands in the genital region. The secretion has a strong musky odour—hence the name Musk-rat. A distance of at least 100 m (300 ft) usually separates one family from the next.

The first litter of young is born early in the spring, in April–May, and there may be up to 3 more litters during the course of the summer and autumn. There are usually 6–8 young in each litter; these are naked and blind, and weigh about 20 g ($\frac{2}{3}$ oz). The young of the first litter may breed during the same summer. Diet consists of the leaves and stems of various aquatic plants; in winter they eat plant roots and some animal food, including bivalves and fish. The extensive digging activities of this species may cause serious damage to river banks, dykes and so on.

69 Mole-rat
Spalax microphthalmus

Identification: Body-length 18·5–27 cm (7¼–10¾ in.), hind-feet 1·9–2·5 cm (⅝–1 in.); weight 140–220 g (5–7¾ oz). These measurements refer to the western form (illustrated); the eastern form is slightly larger. Back grey-brown with a reddish-yellow tinge.

Distribution: South-east Europe from the lowlands of Hungary and Yugoslavia to Bulgaria and eastern Greece, extending up along the Russian coast of the Black Sea to the region round Odessa.

Habitat: Originally associated with steppe regions, but after these had become cultivated it now lives along river banks, the edges of roads and various uncultivated places. It also occurs in fields with crops growing there for several years, e.g. lucerne.

Habits: Lives almost exclusively underground. It is only seen above ground, and then mainly the male, during the mating season in spring. Totally blind, it moves along the ground in a straight line. Tactile and auditory senses are well developed. The underground tunnel systems of this rodent are very extensive and may cover several acres. There is only one individual in each tunnel system but there are probably tunnels connecting with neighbouring systems. The major part of each system lies at a depth of 10–25 cm (4–10 in.) and these can be recognized by rows of long mounds. The tunnels are about 8–9 cm (3–3½ in.) in cross section. Within a limited area 2–4 vertical shafts lead down from the main system to a depth of 1–3 m (3–9 ft), where there is a further system of horizontal tunnels connecting a number of nest chambers, storage chambers and latrines. The nest chamber measures about 20 × 30 cm (8 × 12 in.) and is lined with dry grass. Mole-rats are mainly active from late afternoon until the morning, probably with a couple of intervals for rest. They are formidable diggers, moving the earth partly with the help of the teeth, partly with the head and feet. The head is used to push the loosened soil upwards. Diet consists exclusively of vegetable matter, mainly herbaceous plants and tree roots. In the case of the former they gnaw off the rootlets and then pull the whole plant down into the tunnel by its main root. Stores are collected for winter. Little is known of their breeding habits, but they certainly have only one litter of young per year.

True mice and rats

There are numerous species of the subfamily Murinae in all parts of the Old World. In America there are other rodents, which look similar, but the only species of Murinae in the New World have been introduced by man. Compared with the voles, the true mice have a longer and more pointed muzzle, larger eyes and ears, and longer limbs and tail. The latter is only sparsely haired. As in the voles, there are 3 cheek teeth, but these differ in having roots and lower, 'knobbly' crowns. True mice may be terrestrial or arboreal whereas the voles could be described as largely subterranean.

Coarse vegetable food plays an important role in the diet of most species of Murinae, but a considerable amount of animal food is also eaten and there are some species which are primarily insectivorous.

70 Harvest mouse
Micromys minutus

Identification: The smallest species of rodent in Europe. Body-length 5·8–7·6 cm (2¼–3 in.); tail long, 5·1–7·2 cm (2–2⅞ in.), and with very short hair; weight 6–10 g (up to ⅓ oz). Back pale reddish-brown to yellow-brown, sharply demarcated from the whitish underparts. Eyes and ears relatively small and muzzle rounded and very blunt.

Distribution: Extends from western Europe through Asia to China. Absent from Scandinavia, north Finland, Spain, Portugal, the Alps, south Italy and the southern Balkans. In the British Isles the main areas of distribution are in the south and east of the mainland. It is totally absent from Ireland.

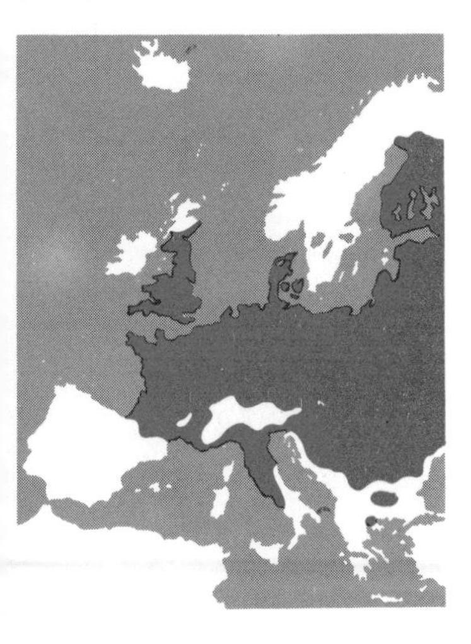

Habitat: Cornfields, especially in fields of oats and wheat, also in dry reed-beds, along hedgerows and other places with long grass or tall-stemmed weeds.

Habits: Activity takes place in daylight and at night, with several periods of rest which are spent in the nest. A remarkably agile climber, ascending thin stems by using the distal, slightly flattened part of the tail as a prehensile organ. On the other hand, it cannot jump well, nor does it run particularly fast. In summer this species lives at some height from the ground in tall vegetation. The nests are built at a height of up to 1 m (3 ft); they are spherical, about 10 cm (4 in.) across, skilfully woven out of blades and stems of grass, and attached to several tall stems. There is a circular entrance at the side. One of the larger nests is lined with plant fibre and used as a breeding nest. The gestation period is about 21 days. The breeding season lasts from April to September and there may be several litters, each with 3–7 young. The latter have their eyes open after about a week and a few days later they start making short excursions from the nest. They are almost adult-size at 3–4 weeks and sexually mature by the fifth week. The life-span is 16–18 months. Several families of Harvest mice often live within a small area. Winter nests are made of moss and are often built in burrows near the surface. In certain parts of their range they are said to store food in underground chambers. They also enter stacks of straw in barns and similar places. Diet includes insects as well as seeds, fruits and buds. This species utters a series of low-pitched, penetrating squeaks.

71 Striped mouse
Apodemus agrarius

Identification: Easily recognized by the distinct black stripe along the back which is reddish-brown (compare with 77). Body-length 9·7–12·2 cm ($3\frac{3}{4}$–$4\frac{3}{4}$ in.); tail relatively short, 6·6–8·8 cm ($2\frac{5}{8}$–$3\frac{1}{2}$ in.), and always shorter than the body; weight 16–25 g ($\frac{1}{2}$–$\frac{7}{8}$ oz). Ears relatively small.

Distribution: A south-eastern form which extends westwards to north Germany and Denmark, but is absent from Scandinavia and the rest of western Europe, including the British Isles. It extends eastwards through Asia to China.

Habitat: Along the edges of woodlands, also in meadows and fields providing there are scattered bushes or trees in the vicinity. Not found in arid areas.

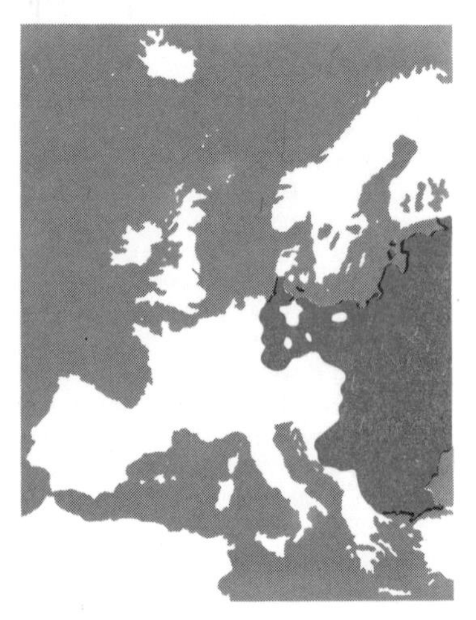

Habits: Very similar to those of the following 2 species, but it is not strictly nocturnal as activity often starts in the afternoon. It is a poor climber and lives mainly on the ground or in burrows. Subterranean tunnels are dug which extend down to a depth of 35–40 cm (13–15 in.). There are 3–4 entrance-holes and a number of chambers for resting and storage. The nest chamber is lined with dry grass. The gestation period is 21–23 days and there are 2–4 litters per year, each with 3–8 young. This species is often found in barns and cellars in winter. Diet is chiefly vegetable food, such as peas and beans, but animal food is also eaten, including considerable quantities of earthworms. The Striped mouse is host to a bacterium which can be transmitted to cattle and pigs, causing serious illness and resulting in premature calving, stillborn piglets and blood in the milk and urine.

72 Wood mouse
Apodemus sylvaticus

Identification: Body-length 7·7–11 cm (3–$4\frac{1}{4}$ in.), usually less than 10 cm (4 in.); tail usually shorter than the body, 6·9–11·5 cm ($2\frac{3}{4}$–$4\frac{1}{2}$ in.); hind-foot seldom over 22 mm ($\frac{7}{8}$ in.) long; weight 14–22 g ($\frac{1}{2}$–$\frac{3}{4}$ oz). Back brown, underparts greyish-white, usually with a transition zone of yellowish fur. The yellow marking on the chest never forms a complete collar. Also known as the Long-tailed field mouse.

Distribution: Most of Europe, including the British Isles and Scandinavia. It extends to North Africa and into

Asia. Its range in the Far East is still subject to a certain amount of controversy as the Far Eastern forms are sometimes treated as separate species.

Habitat: Fields, hedgerows, scrub, heaths and moors, less frequently in dense forest than along the edges and in more open woodland. Also abundant in gardens and around farm buildings. In winter it often enters outhouses and is even found in kitchens and larders.

Habits: Mainly nocturnal. Moves very fast on the ground, making characteristic long leaps—up to 80 cm (31 in.) —and it also climbs and swims well. Each family has a home range but the areas occupied by neighbouring family units overlap with each other. Runways in leaf litter are used and where the soil is suitable complex underground burrows are excavated. Each tunnel system usually has two entrance-holes; at one hole there may be a quantity of loose soil, but none at the other, which suggests that the second hole has been dug from below. Winter nests are built to a depth of 50–100 cm (20–40 in.); the chamber is 15–20 cm (6–8 in.) in diameter and lined with shredded grasses or bits of straw and so on. There are also storage chambers each of which may hold up to 7 litres (12 pints) of corn. Nests are sometimes built in hollow trees or under tree-stumps. Large quantities of seeds, including acorns and beechmast may be stored under the roots. There is a certain amount of foraging activity above ground, even in hard winters. In summer the males and young mice in particular build simpler nests nearer the surface, while the females build breeding nests at the end of a burrow 10–20 cm (4–8 in.) below the surface. Breeding starts in March–April and may continue until December or through the winter in the more southerly parts of the range. The gestation period is about 25 days. There are at least 3 litters per year, each with an average of 5 young. These start to leave the nest after about 2 weeks and are completely independent at an age of 4 weeks. The young of the first litter breed the same summer, but those born in the autumn may not breed until the following spring. They feed primarily on conifer seeds, hazelnuts, beechmast, cereals and other seeds, but also eat considerable amounts of buds, berries, insects and spiders. A fir cone gnawed by a mouse (73c) can be distinguished from one gnawed by a squirrel (44f): the scales are more neatly gnawed off, including the lowermost scales, thus the cone looks less ragged and the base becomes rounded. This species utters a series of high-pitched squeaks.

73 Yellow-necked mouse
Apodemus flavicollis

Identification: Body-length 8·8–13 cm ($3\frac{3}{8}$–$5\frac{1}{8}$ in.) usually over 10 cm (4 in.); tail usually longer than the body, 9·2–13·4 cm ($3\frac{5}{8}$–$5\frac{1}{4}$ in.); weight 22–48 g ($\frac{3}{4}$–$1\frac{3}{4}$ oz). Underparts white and normally sharply demarcated from the brown back. A prominent, transverse yellow spot on the chest between the fore-legs which sometimes forms a complete collar. In adults the hind-feet are at least 23 mm ($\frac{7}{8}$ in.) long.

Distribution: In many areas of Europe but absent from Portugal, Spain, western France and most of Italy. In the British Isles it is absent from Scotland and Ireland; the majority of records come from southern England where the distribution appears to be patchy. In Scandinavia it is found further north than the Wood mouse (72).

Habitat: Prefers mature woodland to forests with shrub layer or more open

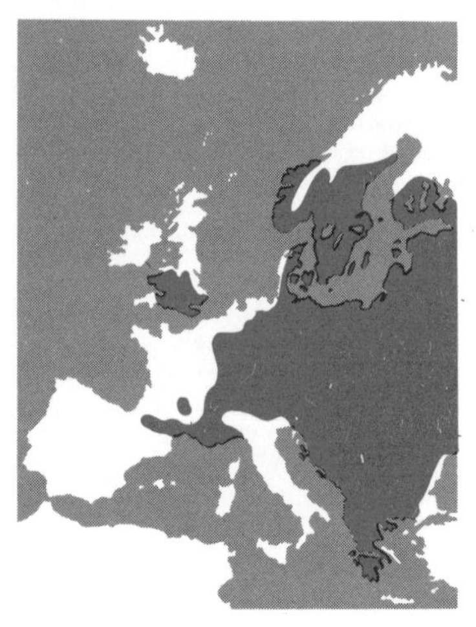

country with scrub and is particularly associated with mature beechwoods. In other respects it is found in similar places to the Wood mouse. The two species overlap to some extent and it is thought that some interbreeding takes place. In winter the Yellow-necked mouse often enters buildings where fruit crops are stored.

Habits: Similar in general to the Wood mouse but tends to be more arboreal and to use existing cavities or burrows, rarely digging extensive tunnel systems.

Rock mouse
Apodemus mystacinus

Identification: Larger and paler than the Wood mouse. Body-length up to 15 cm (6 in.), and tail about the same. Ears very large.

Distribution and habitat: In Europe it occurs only in the southern Balkans among dry scrub on rocky slopes. Also occurs from Asia Minor to Israel.

Habits: Little information available.

74 House mouse
Mus musculus

Identification: Can be distinguished from the Wood mouse (72) and the Yellow-necked mouse (73) by the shorter muzzle, the uniformly coloured tail, the shorter hind-feet, 15–19 mm ($\frac{1}{2}$–$\frac{3}{4}$ in.), and by having a notch at the back of the upper incisors (compare figs. 73b and 74b). Weight 10–24 g ($\frac{1}{3}$–$\frac{7}{8}$ oz). Several forms have been described, of which only 2 will be described here. The darker of these has a body-length of 7·5–10·3 cm (3–4 in.) and a tail-length of 7·2–10·2 cm

($2\frac{7}{8}$–4 in.). The back is grey-brown and the underparts are more or less the same colour but often with a yellowish sheen. The paler form, on the other hand, is typically an outdoor mouse. In summer it lives in fields, waste land, gardens and parks, more rarely in woodland and then only in the vicinity of houses. It enters buildings in winter.

Habits: Mainly nocturnal but some individuals show considerable activity by day. Although many mice may be found living in the same area, the dominant males each live together with one or more females in a territory which is vigorously defended against other members of the species. Where the population density is very high the weaker individuals are unable to establish a territory and will not breed. Normally the House mouse breeds throughout the year but in winter the outdoor populations do so at a much reduced rate. When living in buildings there may be from 6 to 10 litters per year. The gestation period is 19–20 days and there are usually 4–8 young per litter; the period of gestation may be prolonged by about 10 days if the female is still suckling a previous litter. The young become independent after about 18 days and they are sexually mature at an age of about 6 weeks. The life-span is 1–$1\frac{1}{2}$ years. The nest is built in a secluded cavity and made out of almost any kind of soft material that is available. Diet consists mainly of vegetable matter, with some animal food. House mice require very little water. They are good swimmers but do not take readily to water. They are excellent climbers and do considerable damage by gnawing and fouling food stores. A characteristic musky scent usually betrays their presence in a house even if their droppings are not noticed. Field mice do not have this smell and when they enter houses in winter they drive out the House mice.

75 Black rat
Rattus rattus

Identification: Several colour types occur, the commonest being greyish-black. The *alexandrinus* form (75a) is brownish-grey on the back and grey—not pure white—on the underparts. Compared with the next species, the Black rat is more slenderly built: muzzle more pointed, ears larger and thinner, coat shorter and less shaggy. Body-length 15·8–23·5 cm ($6\frac{1}{4}$–$9\frac{1}{4}$ in.); with a longer tail, 18·6–25·2 cm ($7\frac{3}{8}$–10 in.)—longer than in the next species; weight up to 215 g ($7\frac{1}{2}$ oz).

Distribution: Originally a native of south-east Asia, which started to

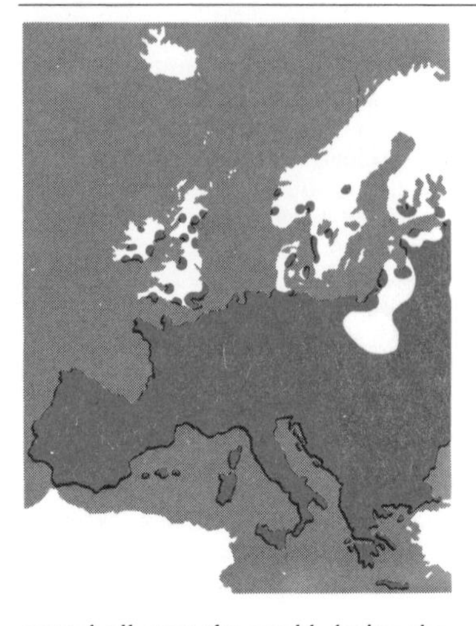

spread all over the world during the Middle Ages, mainly with the help of man. Black rats occur everywhere in Europe, except in the British Isles, Denmark, Scandinavia and Finland where there are only a few populations, mainly in and around large seaports. At one time they were more widespread in these countries but have been driven out by competition from the Brown rat.

Habitat: Principally in warehouses in docks and ports, usually occupying the upper storeys, but also in residential property and on board ships. Although primarily associated with buildings, they are not totally dependent on an urban environment. In some areas they live in trees, usually on cultivated land near farms.

Habits: Mainly active at night. Adept at climbing, and much less terrestrial than the Brown rat. In warehouses various ducts, pipes, cables, rafters and so on are used as runways. This species jumps well, but swims less readily than the Brown rat. The shape and size of rat droppings are indicative of which species is present. In the Black rat these are slightly curved, about 10 mm ($\frac{3}{8}$ in.) long and 2–3 mm (about $\frac{1}{8}$ in.) across, whereas those of the Brown rat are cylindrical and measure about 17 × 6 mm ($\frac{5}{8} \times \frac{1}{4}$ in.). In both species a large number of families share the same general feeding area, keeping to familiar surroundings within a comparatively small compass. Nests are built among stored goods, between rafters, behind partitions, and in cavities of the roof or walls. Where conditions are suitable breeding may continue throughout the year. The gestation period is about 21 days; there are 3–5 litters per year, each with an average of 7 (5–10) young. Sexual maturity is reached at an age of 3–4 months. Black rats are omnivorous and feed on a wide variety of foodstuffs. A preference for vegetable food is shown, particularly for grain, but they will eat almost any form of stored food. It was this species that brought bubonic plague to Europe during the Middle Ages. It can also transmit other dangerous diseases to man and animals.

76 Brown rat
Rattus norvegicus

Identification: More heavily built and coarser in appearance than the Black rat. Muzzle blunter, ears smaller and thicker, coat looks rougher. Body-length 21·4–27·3 cm ($8\frac{1}{2}$–$10\frac{7}{8}$ in.); tail 17·2–22·9 cm ($6\frac{3}{4}$–$9\frac{1}{8}$ in.)—usually shorter than in the preceding species; weight up to 500 g (18 oz). Back brownish-grey, paler on the underparts, but there are several colour variants. Completely black specimens are not particularly rare and the white laboratory rat is an albino form.

Distribution: Originally native to south and east Asia, the Brown rat spread westwards in the seventeenth century, partly by its own migrations and partly on board ships. It is now distributed all over the world. In many places, especially in northern Europe and the British Isles, this species has driven out the Black rat.

Habitat: Mainly associated with inhabited areas, but unlike the Black rat it moves out into more open country during the summer, often living near water. Some live along the seashore but many more take up residence in rubbish dumps, sewers, cellars, the ground floor of warehouses and in farm buildings.

Habits: Mainly terrestrial and an active burrower. Extensive tunnel

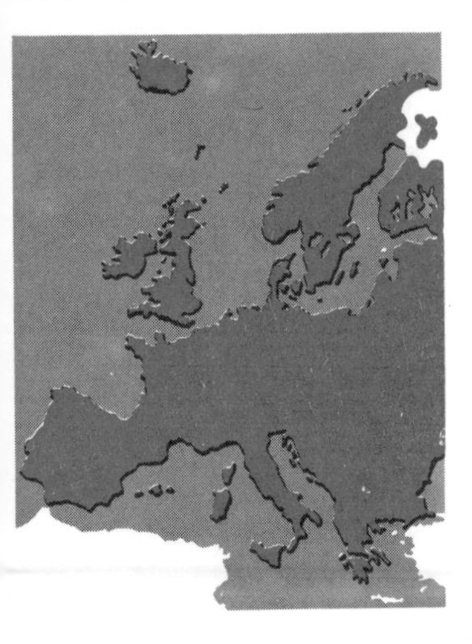

systems are dug to a depth of 40–50 cm (15–20 in.). There are a number of entrance-holes which are connected by runs. Brown rats swim and dive well but are less agile at climbing than the Black rat. They are mainly active at night, particularly at dusk and early in the morning, and if they are seen moving around in daylight this is usually due to hunger or disturbance. They are wary and suspicious of any change in their immediate surroundings. For instance, when a familiar haystack is disturbed the rats all leave. In addition to these small-scale movements, there are frequent reports of a mass exodus in which considerable distances are covered. The Brown rat is extremely aggressive and the females, in particular, defend their young vigorously. Although large numbers of families settle in the same area there is a considerable amount of aggression between members of the same species. When living in sheltered conditions breeding may continue throughout the year. Nests are made out of almost any convenient material; they are built below ground when living out-of-doors and in some kind of cavity when occupying a building. The gestation period is 24 days. The number of young in the litter is correlated with the weight of the mother but is usually 6–10, and there are 3–5 litters in the year. The new-born young weigh about 5 g ($\frac{1}{6}$ oz) and are naked and blind. They leave the nest when they are about 3 weeks old; the females are sexually mature at an age of about 3 months, the males somewhat earlier. The annual mortality of adults is 91–97 per cent. Thus, less than 1 out of 10 survives more than a year. Mortality is even higher among the young when populations are at peak density. Diet is similar to that of the Black rat. When living off the country, as distinct from the warehouse type of environment, the Brown rat eats a comparatively wide range of animal as well as vegetable food. The daily intake of food corresponds to 10–40 g ($\frac{1}{3}$–$1\frac{1}{3}$ oz) of wheat. This species does an enormous amount of economic damage in many countries and also transmits certain diseases.

Birch mice

The family Dipodidae contains some species such as the birch mice which are similar in external appearance to the true mice and others like the jerboas which are desert rodents highly specialized for jumping. There are two species of birch mice in Europe, the remainder being distributed in steppe country in Asia and western Siberia. Birch mice hibernate. They have 4 cheek teeth in each half of the upper jaw, 3 in each half of the lower jaw.

77 Birch mouse
Sicista betulina

Identification: Body-length 5·2–7 cm (2–$2\frac{3}{4}$ in.); tail about half as long again as the body, 7·9–10·6 cm (3–$4\frac{1}{8}$ in.); hind-foot 16–18 mm. ($\frac{5}{8}$–$\frac{3}{4}$ in.); weight 6·5–13 g ($\frac{1}{4}$–$\frac{1}{2}$ oz). Back yellow-brown with a narrow, black dorsal stripe extending from the muzzle to the base of the tail; underparts greyish-white to greyish-yellow without any sharp demarcation of the colours.

Distribution: Only in scattered populations although the range extends from the coasts of the Baltic to Siberia, including parts of Denmark, Scandinavia and Finland, and the mountain regions of central Europe. This scattered distribution suggests that these populations are the remains of what was formerly a much wider range.

Habitat: In open woodlands with plenty of secondary growth or understorey, especially of birch, but in summer is also found in meadows, cornfields and even in marshland.

Habits: Mainly nocturnal, although some activity occurs from time to time during daylight. A very agile climber which moves about high up in trees and bushes. When climbing the tail is used as a support. It also jumps well but rarely enters the water. Birch mice run along the ground, not very fast, with the tail raised. They build spherical summer nests of moss and grass at ground level. In the autumn they prepare a special winter nest in a hole dug in the ground or in a rotten tree-stump. They hibernate for about 6 months of the year, remaining totally

inactive from October to May. Even in summer when the weather is bad they may go to sleep for several days, with a reduced body temperature. The breeding season is May–June and there appears to be only a single litter. The gestation period is 4–5 weeks. Litter-size varies from 2 to 6. The young develop slowly and are 5 weeks old before they leave the nest and become independent. Diet consists of a high proportion of animal food, such as insects and larvae associated with bark and rotting wood, but they also eat buds, seeds and berries. The life-span may be up to 3 years.

Steppe mouse
Sicista subtilis

Identification: About the same size as the preceding species, but the tail is always less than 9 cm ($3\frac{1}{2}$ in.). The hind-foot is also shorter, 13·7–16·5 mm ($\frac{1}{2}$–$\frac{5}{8}$ in.). The black dorsal stripe is flanked by 2 stripes which are paler than the rest of the back.

Distribution and habitat: Principally associated with steppe and semi-arid areas. The range extends from Rumania in the west, with scattered populations in Hungary and Austria, to the southern areas of Russia eastwards into Asia.

Habits: Similar to those of the preceding species. Diet includes seeds and insects associated with grassland, and grain in cultivated areas.

Porcupines

78 Crested porcupine
Hystrix cristata

Identification: Body-length 57–68 cm ($22\frac{1}{2}$–$26\frac{1}{2}$ in.); tail 5–12 cm (2–$4\frac{1}{2}$ in.); weight up to 12 kg (26 lb). The rear half of the body, including the back, sides and tail, has a large number of thick spines, each up to 40 cm (15 in.) long; the spines on the front half of the upper parts are thinner but form a dense covering. The spines are banded in black, dark brown and white. The underparts have coarse, black fur.

Distribution: Restricted in Europe to south Italy, Sicily and parts of the Balkans, but the range of this species extends into northern Africa and Asia Minor.

Habitat: Dry slopes with bushes and trees, including open woodland, usually in the vicinity of cultivated land.

Habits: A nocturnal animal which spends the day in subterranean holes, usually in some kind of natural cavity but sometimes in a burrow it has dug for itself. More than one porcupine may occupy the same burrow but it seems likely that they are members of a family unit. They emerge in the evening to search for food. Their diet is primarily vegetarian although they are known to eat carrion. Fruit, bark and roots provide the chief items, and they do a considerable amount of damage to agricultural crops in certain areas.

When alarmed a porcupine will stop abruptly in its tracks, with its back to the enemy and erect the sharp spines so that these face in all directions; at the same time it rattles the hollow quills on the tail and makes a continuous grunting sound. Thus the total effect is decidedly impressive even though the porcupine is not really an aggressive animal. The spines break off very easily and become embedded in the enemy's flesh where extraction is difficult.

Details of the reproductive cycle are not yet fully known but mating takes place in the spring and it is possible that there is only one litter per year. The gestation period is about 3 months and the litter-size is 1–4. The young are born in a nest below ground and are well developed at birth. They are covered with spines which harden after a week or two. Porcupines have been known to live for up to 20 years.

Coypu

79 Coypu
Myocastor coypus

Identification: Body-length 42–60 cm (16½–24 in.); tail 30–45 cm (12–15 in.); weight 6–9 kg (13–19 lb). Distinguished from the Beaver (48) and the Musk-rat (68) by the rat-like tail which is rounded and tapers gradually towards the tip. The feet are webbed. The colour of the back varies from yellowish-grey to almost black, the underparts being pale grey.

Distribution: A native of South America, whence it has been introduced as a commercial fur animal into a number of European countries, including the U.S.S.R. and the temperate parts of Asia. Escaped specimens have established feral populations in many countries, including England.

Habitat: The banks of lakes and slow-flowing rivers with marsh vegetation.

Habits: Mainly active at dusk and dawn, but also during the night, and is occasionally seen in daylight. The numbers living in one area are often high but each individual is solitary. Coypus dig tunnels and holes in banks, but also build nests of plant material above ground. There are also conspicuous runs through reed-beds and other marsh vegetation. They move very slowly and clumsily on land but dive and swim well. They breed throughout the year. There is a long gestation period, about 130 days. There may be 2 litters per year, each with an average of 5 young. These are born fully furred and with their eyes open; within a few hours of birth they can move around actively, even entering the water, and are capable of eating solid food when only a day old. The female is able to suckle the young while swimming at the surface; the teats are positioned so high on the sides that they are above the level of the water. Coypus feed chiefly on various aquatic and marsh plants, but they will also eat freshwater mussels. In some areas they do considerable damage to agriculture. In East Anglia, for instance, there has been an intensive campaign to reduce their numbers. Coypu fur is known commercially as nutria.

Carnivores

The order Carnivora contains about 300 species, distributed in all parts of the world. The majority live in the tropical regions and only a few in the true arctic. They vary considerably in size: at one end of the scale, for example, there is the very small Weasel and at the other end there are very large animals, such as the bears, but the majority are medium-sized. External appearance is less varied: most carnivores are somewhat elongated and slender in build, often very sinuous; none, however, can be described as highly specialized in one particular direction. Nearly all carnivores will swim readily and a few are mainly aquatic. The remainder are primarily terrestrial, some having a considerable turn of speed, others being very agile climbers. All have dense fur, made up of woolly and guard hairs, which in some species form a thick, soft pelt. The quality of the pelt has made them commercially desirable, so much so that in recent years quite a few species have been threatened with total extermination.

The most characteristic structural character of the carnivores is the dentition with its highly specialized functional differentiation. The most primitive dentition is found in the dog family (Canidae) in which each half of the upper jaw has 3 incisors, 1 canine, 4 premolars and 3 molars; the lower jaw has the same complement of teeth except that there are only 2 molars. The other carnivore families show various reductions, partly of the premolars, partly of the molars; the latter condition is most marked in the cat family (Felidae). The extent of this reduction and the form of the cheek teeth is closely correlated with the type of food. The incisors, usually 3 pairs in the upper and in the lower jaw, are small and chisel-shaped. In most carnivores the canines are large, pointed and slightly curved. The premolars and molars are usually compressed, particularly the most anterior. The molars are relatively low and broad, with a rough biting surface. The last upper premolar and the first lower molar on each side of the jaw have cutting edges which bite against the edge of the opposed tooth; these teeth are known as the carnassials. The jaw articulation only allows an up and down movement and the jaw musculature is usually extremely powerful, so that the dentition as a whole provides a very efficient organ for biting, shearing and crushing. On the other hand, the teeth are incapable of true mastication.

Various sense organs are well developed, particularly smell and hearing, but also the tactile sense which is linked with the vibrissae or whiskers. The limbs vary considerably in form but have a minimum of four claws on the digits of both fore- and hind-feet; the first digit or thumb in the majority of species has completely disappeared on the hind-foot and is more or less reduced on the fore-foot. Some carnivores are plantigrade, treading on the whole foot, whereas others walk only on the tips of the toes; the latter include all those that are fast runners. All carnivores have scent glands; these are located in many different areas of the body but the commonest site is near the anus. The secretion of scent is used, in general, as some kind of communication: among other things territory is marked with it but in some species the smell is unpleasant so that the secretion serves also as a highly efficient means of defence. Carnivores which use scent in defence usually have a very striking colour pattern.

Most carnivores feed on animal food and correlated with this their gut is short and uncomplicated. A certain amount of plant food is also eaten by many of them and a small number are entirely vegetarian. Longevity and fecundity vary, but the larger species are long-lived and have a low fecundity. A protracted period of gestation is another feature of some carnivores: after mating takes place the fertilized egg virtually stops growing—a process known as delayed implantation—and development of the embryo is resumed only after a certain period of time.

In Europe the order Carnivora is represented by 6 families: the Viverridae (civets and mongooses), Felidae (cats), the Mustelidae (Stoat, Weasel, Mink, etc.), the Canidae (dogs), the Ursidae (bears) and the Procyonidae (raccoons).

Civets and mongooses

Members of the Viverridae are relatively small, elongated animals with rather sharp features, short legs and a long tail. Scent glands are present near the anus. They have 32–40 teeth. There are only 2 species in Europe but the family is represented in the Old World by about 75 species.

80 European genet
Genetta genetta

Identification: A relatively small carnivore resembling a cat, but with a slenderer body, a more pointed muzzle, shorter legs and a longer neck and tail. Body-length 47–58 cm (18–24 in.); tail 41–48 cm (16–20 in.), long and pointed; shoulder-height 18–20 cm (7–8 in.); weight around 2 kg (4¼ lb). The fur which is short and smooth is greyish or yellowish with longitudinal rows of dark spots. The tail has 9–10 dark rings and is black at the tip.

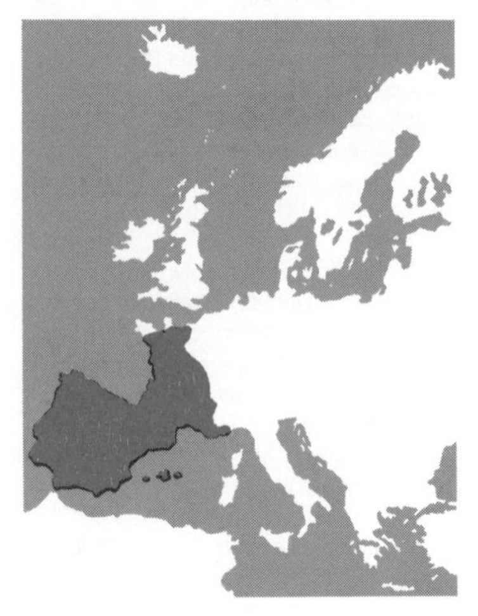

Distribution: In Europe, the Iberian Peninsula and western France eastwards to a line running from Normandy to Provence. Now and then a few individuals are found in more northerly and easterly regions, including Belgium and north-east France. The main area of distribution is in Africa.

Habitat: Damp forests in mountain regions and along rivers in heavily wooded lowland areas.

Habits: Hunts exclusively by night. Moves extremely quietly in the terrain, slinking along through low herbage and bushes, using every natural depression and bit of cover. Kills all small animals, such as mice and other rodents and eats them immediately; birds and their eggs, insects and other invertebrates are also included in the diet. Swims remarkably well, climbs with agility and attacks birds roosting in trees as well as birds' nests; poultry runs and dovecotes are also raided. In spite of this harmful activity the genet is regarded as useful to man because it kills enormous numbers of rats and mice. In certain districts it is even kept as a domestic animal with this end in view.

Genets spend the day in a den, situated in a well-hidden place, often between bushes or rocks or in a hollow tree. There are 2 litters in the year, each with 2–3 young. This animal has a strong smell and marks its territory by leaving heaps of faeces at certain well-defined places.

81 Egyptian mongoose
Herpestes ichneumon

Identification: Body-length 51–55 cm (20–21½ in.); tail 33–45 cm (13–17 in.), long and very pointed; shoulder-height 19–21 cm (7¼–8¼ in.), weight 7–8 kg (15–17 lb). Legs short and almost hidden in the long fur during locomotion; ears short and broad. The fur consists of short, yellowish, woolly underfur and stiff, black and yellowish guard hairs, 6–7 cm (2–2¾ in.) long, giving the whole animal a grey speckled appearance.

Distribution and habitat: A native of many parts of Africa, which also occurs in the south of Spain and in Italy and Yugoslavia.

Habits: Hunts by day and night, but chiefly nocturnal. Feeds on a wide range of animal food which includes insects, earthworms, lizards, snakes, birds and small mammals. Often attacks domestic fowl and other poultry. Digs burrows in the ground in which the female gives birth to 2–4 young in April–May. As these grow up they accompany the mother while she is hunting. The young remain in the family group until shortly after the female has produced her next litter in the following spring. It is typical of the members of this family that, when hunting, they move in a long line following one behind the other.

Cats

Members of the cat family, Felidae, have relatively long legs, a powerfully built body and a rounded head. The claws are retractible. The cheek teeth behind the carnassials are very much reduced.

There are very few species in Europe, indeed some authorities consider that there are only two (the Lynx and the Wild cat), while others prefer to recognize the southern form of lynx as an additional species.

82 European lynx
Felis lynx

Identification: Body-length 80–130 cm (31–50 in.); tail short and broad, 11–25 cm (4¼–9 in.), with a black tip; shoulder-height 60–70 cm (23–27 in.); weight 14–30 kg (30–65 lb). Legs long and powerful; ears with a tuft of long, black hair. Fur very short, the ground colour reddish-grey in summer, greyish-white in winter, with varying amounts of dark spots.

Distribution: Northern and eastern Europe, eastwards through northern Asia to the Pacific Ocean. Occurs in small numbers in Sweden—only a few hundred individuals—and in eastern Finland.

Habitat: Mainly in dense, tall stands of mature trees, preferably in forests, with rocky outcrops where many of the trees are blown down by the wind. In the mountains of Scandinavia also in more open birch woods, and in southern Sweden it has become adapted to some extent to living in cultivated areas.

Habits: Hunts mainly at dusk and during the night. Prey is captured partly by stalking and partly by lying in ambush on a rocky ledge or up a tree. Usually moves rather slowly, relying on a stealthy approach to prey but when pursued it can travel fast, although only for short distances. Climbs and swims well. Prey consists of small rodents, hares, squirrels, foxes, small ungulates, large and small birds and fish. Sometimes domestic cats and dogs are also killed.

Mating takes place in February–March, and at this time the call of the males can be heard from a distance: starting as a high-pitched howl, it ends in a low moan. After a gestation period of about 9 weeks, the female produces 2–3 young in April–May. These are born in a lair in a dry, well-sheltered place, either on the ground in undergrowth, under rocks or fallen trees, or sometimes in a hollow tree. The young are suckled until they are about 6 months old, but shortly after birth the female also brings them small prey as supplementary diet. The male takes no part in rearing the young. Gradually as the young grow up they accompany the female on hunting trips; they do not become fully independent until January–February of the following year. This species was formerly much hunted, mainly for the valuable fur.

83 Spanish lynx
Felis pardina

Identification: Very similar to the preceding species, but usually smaller and with a closer pattern of large and small spots. Body-length 85–110 cm ($33\frac{1}{2}$–43 in.); tail 12–13 cm ($4\frac{3}{4}$–5 in.). Possibly not a valid species, but only a subspecies of *F. lynx*.

Distribution: Formerly common in many parts of southern Europe, but now exterminated in many areas, including France and Italy. It is now commonest in south-west Spain but also occurs in the Balkans and Carpathians.

Habitat: Partly in mountain forest, partly in scrub in low-lying plains, as in the Coto Donana in south-west Spain, where the largest population—about 150 individuals—is found.

Habits: Little known, but in general probably very similar to those of the preceding species. Has been much persecuted by man.

84 Wild cat
Felis silvestris

Identification: Considerably larger and more sturdily built than the domestic cat; the body has well-marked dark and light transverse strips, which may be somewhat similar to those of the domestic tabby cat. Body-length 47–80 cm ($18\frac{1}{2}$–31 in.); tail 26–37 cm ($10\frac{1}{4}$–$14\frac{1}{2}$ in.), relatively shorter than in the domestic cat but much more bushy and rounded at the tip; shoulder-height 35–40 cm ($13\frac{1}{2}$–$15\frac{1}{2}$ in.); weight 5–10 kg (11–22 lb).

Distribution: Central, south and south-east Europe, but occurs only in small numbers. Also in northern Scotland, but absent from Scandinavia and other parts of north Europe.

Habitat: Large, continuous areas of forest and scrub, mainly in inaccessible mountain regions. Hunting territories include open ground.

Habits: Mainly active at night, particularly at dusk and dawn. Spends the day lying up in favourite places, such as on a wind-blown tree or rocky ledge, where it rests, sleeps or basks in the sun. In summer the Wild cat may also hunt during the day; it is most frequently seen in the autumn when busy feeding up for the winter. Usually solitary but sometimes a pair is seen hunting together although never when the female has kittens, as the male is liable to attack them. The home range consists of some 60–70 hectares (150–170 acres), intersected by a number of paths. Territory is marked by scent secreted from glands in the region of the anus. The droppings are left lying on the ground and not buried as in the domesticated cat. Prey consists mainly of small rodents, but larger mammals such as hares, rabbits and squirrels are also taken. Wild cats also eat birds, fish and insects.

Mating takes place on the Continent in February–March—in Scotland from early March—and after a gestation period of about 9 weeks the female gives birth to a litter of 2–4 kittens. These are born in a well-hidden place among rocks, under fallen trees or in large birds' nests. When the kittens are 10–12 weeks old they begin to accompany the mother on hunting trips. At an age of about 5 months the young leave the mother and are fully grown at 10 months. Sometimes more than one litter per year is produced by a female but it is thought that this may be due to some admixture of ancestry; it is known that the true Wild cat will mate with feral domestic cats. The calls of the Wild cat are similar to those of the domestic cat, but the voice is louder and more raucous.

Mustelids

The mustelids or Mustelidae are mostly fairly small mammals with an elongated body and short legs. The claws are not retractable. In other respects it is difficult to give a comprehensive definition to cover all the species. As in the cats there is a reduction of the cheek teeth behind the carnassials. There are 30–38 teeth. Some mustelids have delayed implantation. Scent glands, particularly in the anal region, are usually well developed.

85 Pine marten
Martes martes

Identification: Body-length 42–52 cm ($16\frac{1}{2}$–$20\frac{1}{2}$ in.); tail 22–26·5 cm ($8\frac{1}{2}$–$10\frac{1}{2}$ in.); shoulder-height up to 15 cm ($5\frac{1}{2}$ in.); weight 1–1·8 kg ($2\frac{1}{4}$–4 lb). Fur chestnut-brown to almost black with a yellowish throat patch of varying extent (85a). Ears relatively long and broad, pale at the edges. Tail with relatively long fur. Muzzle dark brown. In winter the feet have dense fur on the underside: this makes their tracks in the snow look rather woolly since definition of the pad mark is poor (85c). The most posterior cheek tooth in the upper jaw has no indentation on the outer side (85b).

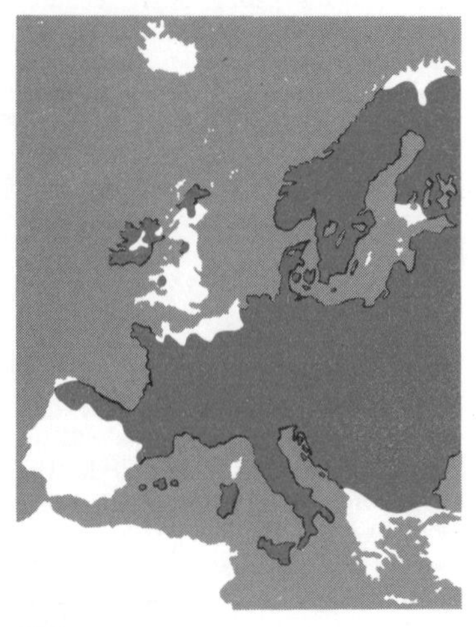

Distribution: Throughout most of Europe—including the British Isles—southwards to north Spain, Italy and Yugoslavia. The range tends to be more northerly than that of the next species but there is considerable overlap. In the British Isles there are only scattered populations and the species is absent from northernmost Scandinavia (above the tree-line). Extends eastwards through the Soviet Union to western Siberia.

Habitat: Mainly mature conifer and mixed forest, but also occurs in deciduous woodlands. Tends to avoid cultivated land, but in the British Isles it is found on open ground as well as in forests.

Habits: Active both by day and night, but mainly at dusk. A very shy and suspicious animal. Moves along the ground in a series of bounds, thus the tracks left by the fore- and hind-feet tend to be close together (85d). The distance between each group of tracks is generally 40–60 cm (15–24 in.), but much longer jumps can be made. Remarkably agile in trees, its climbing and jumping ability enable it to catch squirrels. Resting places are in hollow trees, piles of brushwood or in deserted nests of crows and birds of prey, sometimes even in holes in the ground.

Mating takes place in July–August, but the embryos do not start to develop until January in the following year (delayed implantation). The 2–5 young are born in a warm, lined den in March–April; there is only one litter per year. The young leave the den when they are 7 weeks old, but remain with the mother for some time. They become sexually mature at the age of 2–3 years.

Pine martens hunt on the ground and in trees, and squirrels are an important part of their diet. They also feed on other small rodents, birds and their eggs, and some invertebrates; in autumn a considerable amount of berries and other fruits form part of their diet. The daily requirement of food is about 100–200 g (3½–7 oz). A variety of vocal sounds are made, including various chattering calls and moaning sounds; during the mating season, screeches and growls are heard, somewhat reminiscent of cats fighting. Territory is marked with scent from the anal glands.

This species has been much hunted in Europe for the sake of its valuable fur.

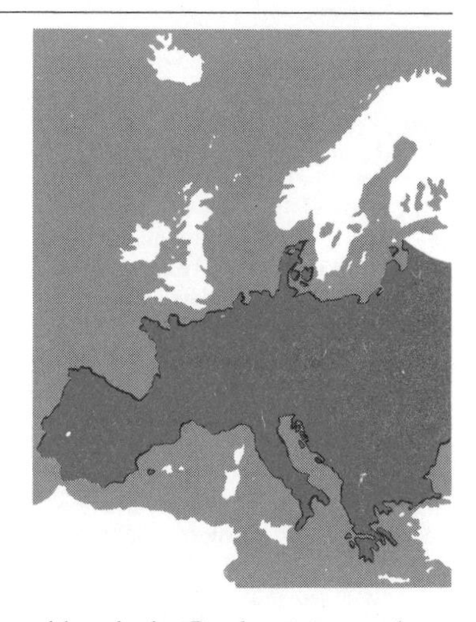

86 Beech marten
Martes foina

Identification: Very similar to the preceding species but the throat patch is white—not yellow—and often split into two areas which extend down the fore-legs (86a); it also has smaller ears and appears shorter in the leg. Body-length 42–48 cm (16½–19 in.); tail relatively thinner and shorter, 23–26 cm (9–10¼ in.); shoulder-height about 12 cm (4¾ in.). Muzzle pale. In winter the undersides of the feet are less densely furred, thus the tracks in the snow are better defined (86c) than in the preceding species. The most posterior cheek tooth in the upper jaw has an indentation on the outer side (86b).

Distribution: Throughout Europe except the British Isles, Scandinavia and Finland. Also absent from many islands in the Mediterranean. Range overlaps with that of the Pine marten although the Beech marten tends to have a more southerly distribution.

Habitat: In contrast to the Pine marten, this species tends to frequent cultivated and inhabited areas. Mainly in and around deciduous woodlands, and in the Mediterranean area on rocky slopes.

Habits: Much the same as those of the Pine marten but more nocturnal, possibly due to greater proximity to human habitations. Resting places are well hidden and include sites in the roofs of outhouses, under the floors of summerhouses, in barns, hollow trees, brushwood, birds' nests and holes in the ground. They catch relatively more rats and mice than the Pine marten and are therefore considered beneficial to some extent. On the other hand they also raid poultry runs and dovecotes. They are less hunted than the Pine marten as their fur has not the same commercial value.

87 Stoat
Mustela erminea

Identification: Body-length 22–29 cm ($8\frac{3}{4}$–$11\frac{1}{4}$ in.); tail 8–12 cm (3–$4\frac{1}{2}$ in.); hind-foot 34–38 mm ($1\frac{1}{4}$–$1\frac{1}{2}$ in.); ear 17–24 mm ($\frac{5}{8}$–1 in.); weight on average 210 g ($7\frac{1}{4}$ oz) for the male, 140 g (5 oz) for the female. Fur relatively sparse and smooth, the hairs at the tip of the tail 40–60 mm ($1\frac{1}{2}$–$2\frac{3}{8}$ in.) long. In summer, upperside red-brown to yellow-brown, underside yellowish-white; in winter, completely white except the tip of the tail which remains black both in summer and winter (87b). This seasonal change of colours occurs only in northern and central Europe, the amount of whitening is variable in the more southern areas of the range (including Britain) and in some regions the Stoat remains brown throughout the year.

Distribution: Throughout Europe (including the British Isles) except in the south. Also in north Asia and North America.

Habitat: In both woodland and open ground, at varying altitudes, the sole requirement being some kind of cover; tends to remain on more open ground in the summer. Often hunts in and along stone walls, and is also seen near buildings in winter.

Habits: Mainly active at night, but also by day in summer when there are young. Bounds along the ground, like the martens, rising on its hind-legs every now and again to have a look around. An excellent climber, although most activity takes place on the ground. Swims well and readily, as for instance when chasing prey. Runways under the snow are used during severe weather and in the arctic part of its range. Several kinds of track are made (87d). Distance between the individual groups of footprints can vary from 20 to 100 cm (8–39 in.); under normal circumstances the average is 50–60 cm (19–23 in.) for the male and about 30 cm (12 in.) for the female. There are no special runways used when hunting; this often takes place along hedgerows and banks. Has a home range of up to 30–40 hectares (75–100 acres) but only a part of this is searched for prey every day. A strong-smelling secretion from a pair of small anal glands is produced when the animal is alarmed and it is possible that scent is also used to mark territory. Prey is tracked down by scent and is pursued both above and below ground.

Stoats are regarded as beneficial to man in so far as they catch many small rodents, moles, rabbits and hares; they also take birds and their eggs, including game-birds, and will sometimes attack domestic poultry. The prey is killed by a bite at the back of the neck but Stoats do not suck blood as is often stated.

The two sexes usually live completely apart except during courtship and mating. Fertilization takes place in late spring or early summer, but there is delayed implantation: the actual development of the embryo does not begin until the following spring and takes only 21–28 days. The complete gestation period is therefore about 280 days. The nest for the litter is in a stone wall, a hollow tree or an underground burrow, dug originally by rats or moles; it is lined with feathers, hair, leaves and so on. There are usually 3–7 young in a litter, although as many as 13 have been recorded, and the young are suckled for about 5 weeks. During the autumn family parties can be seen hunting together. The females mate in their first summer, but the males are not sexually mature until the following summer. Stoat populations fluctuate widely and these fluctuations are related to changes in the numbers of rodents.

88 Weasel

Mustela nivalis

Identification: Body-length 16–23 cm ($6\frac{1}{4}$–9 in.); tail 4–6·5 cm ($1\frac{1}{2}$–$2\frac{1}{2}$ in.); hind-foot 21–35 mm ($\frac{3}{4}$–$1\frac{3}{8}$ in.); ear 12–16 mm ($\frac{1}{2}$–$\frac{5}{8}$ in.); weight on average 85 g (3 oz) for males, 50 g ($1\frac{3}{4}$ oz) for females. Fur short, the hairs on the tail tip 10–16 mm ($\frac{3}{8}$–$\frac{5}{8}$ in.) long. In summer, upperside rich brown, underside pure white; the demarcation line between the two is not sharp as in the Stoat, but irregular and broken. A brown patch often visible at the angle of the mouth. Feet mainly brown. Tail

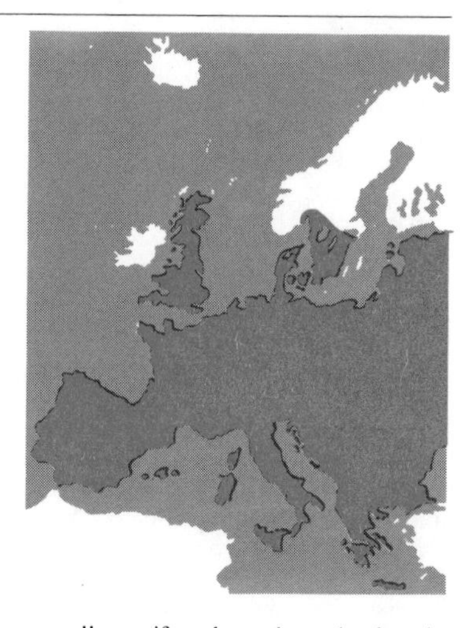

usually uniformly coloured, the tip never has more than a few black hairs at the most. In certain parts of northern Europe Weasels become completely white in winter but in Britain the Weasel is less likely to whiten than the Stoat.

Distribution: Throughout the greater part of Europe, but absent from Ireland and Finland, and in Scandinavia only found south of a line from Oslo to Stockholm. Also in north Asia and northern Africa.

Habitat: In woodland and on open ground, at varying altitudes, also close to buildings, but only in places where there are plenty of voles and mice. Appears to compete unfavourably with the Stoat when under pressure.

Habits: Much the same as those of the Stoat, but on account of smaller size the Weasel can follow mice and voles along their burrows and it is also more subterranean during winter. Hunts both by night and day. The distance between each group of footprints is only 20–30 cm (8–12 in.). The home range is much smaller than that of the Stoat, maximum being 1·7 hectares ($2\frac{3}{4}$ acres). During a night's hunting a distance of 1–2 km ($\frac{2}{3}$–$1\frac{1}{4}$ miles) may be covered. In addition to mice and voles which form the major part of their diet, leverets, small birds and their eggs, invertebrates and so on are also eaten, but poultry is rarely attacked.

Mating takes place from March to August and the gestation period is about 6 weeks. (There is no delayed implantation as in the Stoat.) There are often 2 litters in a season, the first in April–May, the second in July–August. The number of young per litter is usually about 7, but it can vary from 4 to 11. They are suckled for 4–5 weeks. The young of the spring litter become sexually mature during the same summer, but those of the summer litter only start to breed in the following spring.

89 Least weasel
Mustela rixosa

Identification: Body-length 13–19·5 cm (5–$7\frac{1}{2}$ in.); tail 2·8–5·2 cm (1–2 in.); hind-foot 17–30 mm ($\frac{5}{8}$–$1\frac{1}{8}$ in.); ear 9–13·5 mm ($\frac{3}{8}$–$\frac{1}{2}$ in.). Summer coat as in the preceding species but the demarcation line between back and belly is straight and sharp, and there is no brown patch behind the angle of the mouth. More white on the feet, particularly at the sides, than in the preceding species. Fur completely white in winter. Weasels tend to vary in size and markings and some authorities do not recognize the Least weasel as a separate species.

Distribution: Weasels found in Scandinavia north of the mid-Swedish lake region have sometimes been assigned to this species. Other similar populations occur in Finland and eastwards through Russia and northern Asia to north America.

Habitat and habits: As for the Weasel.

90 European polecat
Mustela putorius

Identification: Body-length 32–45 cm ($12\frac{1}{2}$–$17\frac{1}{2}$ in.); tail 13–19 cm (5–$7\frac{1}{4}$ in.); weight up to 1,500 g (52 oz) in the male, 800 g (30 oz) in the female. Fur very long. Upperside paler than underside, palest on the flanks where brownish-black guard hairs only partly cover the yellowish woolly underfur. Tail black with long fur. Ears edged with white. Tracks: see fig. 90b.

Distribution: In most areas of Europe, but absent from the Mediterranean islands and southern Balkans. For-

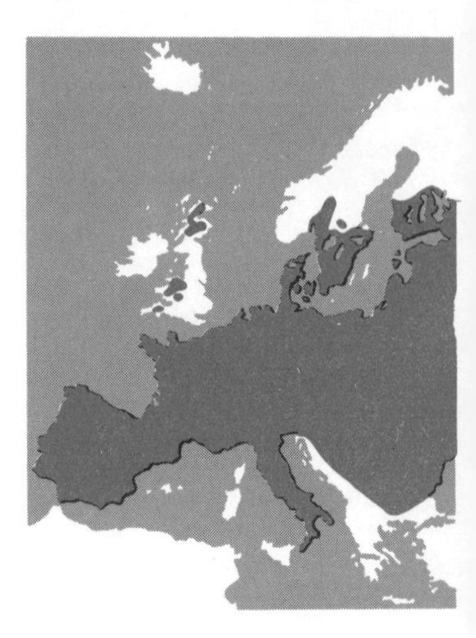

merly widespread in Britain, but now restricted to Wales and parts of Scotland. In Scandinavia northwards to the mid-Swedish lake district and Gudbrandsdal. In Finland only in the south and east.

Habitat: Mainly in thickets and woodland, also in scrub and more open country with rocks; in some areas near human habitation, usually in hedgerows around fields.

Habits: Hunts almost exclusively at night, and spends the day in a den. Prey is located mainly by hearing and scent, in contrast to the martens which hunt mainly by sight. A less agile climber than the martens and spends more time on the ground where it proceeds in a series of bounds (see fig. 90c). Swims well. Hunts rabbits in their burrows. Small mammals form a large part of their diet, particularly mice in winter; in summer mainly birds, reptiles, frogs and fish. Sometimes creates havoc in poultry runs.

The den may be in a hollow tree, under a tree-stump or in a stone wall, sometimes in a hay-loft, or in a burrow dug out specially and side tunnels of disused fox's earths are also used. The den is lined with grass and moss. Mating takes place in March–April, and after a gestation period of 41–42 days the female gives birth to 4–6 young. These are naked and blind and weigh about 10 g ($\frac{1}{3}$ oz). They are weaned by the age of 2 months but may accompany the mother for a few more months. The young Polecats become sexually mature in the following spring. There may be more than one litter per year. Polecats make a variety of sounds, including chattering and hissing. A very unpleasant smelling secretion is produced from a pair of small anal glands when the animal is alarmed. This strong-smelling secretion is probably also used to mark territory. Life-span 8–10 years.

Asiatic polecat
Mustela eversmanni

Identification: Paler than the European polecat (90), ears completely white.

Distribution: In the steppe country of south-east Europe, northwards to Hungary and Czechoslovakia.

Habitat and habits: Lives in completely open terrain, where the burrows of hamsters and ground squirrels may be used, but it also digs burrows of its own. Feeds principally on various rodents and has similar habits, in general, to the European polecat.

The Ferret which is a whitish form of polecat was probably derived originally from *M. eversmanni.* In certain parts of south-east Europe whitish Ferrets live wild in much the same habitat as *M. eversmanni*, and their habits are similar. Ferrets are kept in captivity in many parts of Europe where they are used to force rabbits and rats out into the open, so that they can be killed more easily. These domesticated Ferrets often escape and live on in the wild. Where matings occur between feral Ferrets and wild Polecats, dark hybrids result and these animals are known as polecat-ferrets.

91 European mink
Mustela lutreola

Identification: Body-length 28–40 cm (11–15 in.); tail 12–15 cm ($4\frac{3}{4}$–$5\frac{3}{4}$ in.); weight up to 900 g (30 oz) for males, 600 g (20 oz) for females. Fur short and dark brown, with a varying amount of white on the upper and lower lips. Webbed between the toes, particularly well developed on the hind-feet.

Distribution: Mainly in eastern Europe, extending northwards to about 65°N, including southern and central Finland. Absent from Norway, Sweden and Denmark. Elsewhere there are scattered populations in Poland, eastern Germany, Czechoslovakia, Hungary, Rumania, Yugoslavia, northern and western France.

Habitat: Closely associated with water. Lives mainly in wooded marshland, with plenty of low ground cover, near small streams, and also along the banks of larger rivers, lakes and ponds, where it moves about in marsh vegetation.

Habits: A crepuscular and nocturnal animal, which is also active by day during rainy weather. Moves along the ground in a series of bounds. The distance between each group of footprints is 40–70 cm (15–27 in.). The tracks resemble those of the Polecat (90b). Swims and dives extremely well, capable of remaining submerged for up to 2 minutes. Hunts mainly for small mammals, including Water voles. Also eats fish, crustaceans, frogs and toads, insects and berries; birds are also attacked, including ducks in moult.

The den is always close to water, either in a hollow towards the base of a tree, or under the roots; also digs its own burrow or enlarges those of Water voles and uses natural cavities in banks. An open den is sometimes occupied in a reed-bed. Mating takes place in February–April and the 2–7 young are born in May–June after a gestation period of about 9 weeks. These are tended by the mother for about 10 weeks. The young become fully independent by the end of August and are sexually mature in the following year. During winter they live close to fast-flowing rivers which do not freeze over. The European mink was at one time very common in western and central Europe, but in many areas it has now been completely or partly exterminated due to the value of the fur.

American mink
Mustela vison

Identification: Closely related to the preceding species, which it resembles except for the absence of white on the upper lip; in general, it is somewhat larger and heavier than the European mink. Body-length up to 45 cm (17½ in.); tail up to 21 cm (8¼ in.). The coloration of the head is shown in fig. 91b.

Distribution: Originally native to North America, whence it was introduced into Europe during the first half of the nineteenth century, with a view to breeding on fur farms. The fur is superior in quality to that of the European mink. In many areas—including parts of Britain, Germany, Scandinavia and Finland—escaped individuals have established feral populations. It has also been introduced deliberately in certain places and appears to thrive, as it spreads remarkably quickly.

Habitat and habits: The behaviour and life cycle of feral populations do not differ significantly from those of the European mink.

Marbled polecat
Vormela peregusna

Identification: About the same size as the Polecat (90). Back yellowish with irregular, reddish and brown mottling, and head with a sharply defined pattern of yellow and black. Fur on the tail longer than in the Polecat and the ears are somewhat larger.

Distribution: The steppe and semi-arid regions in the southern parts of the Soviet Union and the adjacent areas of Bulgaria and Rumania.

Habitat and habits: In areas with rocks and scattered scrub. The den is underground, the burrows of large rodents are often used. Feeds mainly on small rodents, particularly gerbils and ground squirrels.

92 Otter
Lutra lutra

Identification: Body-length 50–95 cm (19–37 in.) with thick, brownish fur; tail 26–55 cm ($10\frac{1}{4}$–$21\frac{1}{2}$ in.) long, broad at the root and tapering gradually, with short thick fur; shoulder-height about 30 cm (11 in.); weight up to 15 kg (33 lb), but usually 6–12 kg (13–26 lb). Characteristic spiky appearance as animal emerges from water is due to matting together of guard hairs. Somewhat paler on chin and underpart of neck. Head broad and flat. Ears very small, partly hidden in the fur; when under water the ear can be closed by a fold of skin. Both fore- and hind-feet have a broad web between the toes.

Distribution: Throughout practically the whole of Europe—including the British Isles—extending eastwards into large areas of Asia. Rare, however, in many regions and completely absent from certain islands in the Mediterranean.

Habitat: Lives close to water, along streams, rivers and lakes, in marshland, also along the sea coast and in estuaries.

Habits: Few people have seen otters in the wild but evidence of their presence can often be detected from the remains of half-eaten prey left lying on some prominence, such as a rock or hummock of earth. Droppings are also deposited on raised sites. The faeces—known as spraints—have a distinctive odour. Otters have a number of scent glands and scent undoubtedly has significance in marking territory. Sounds of Otters in the locality are also heard at dusk and after dusk: splashing and various calls, including a characteristic whistling.

In secluded places Otters will emerge during daylight but they are mainly nocturnal. The day is usually spent resting in some kind of hole. The den or 'holt' is often excavated in a bank and there is nearly always a tunnel leading directly under water; natural cavities under tree-roots, in hollow trees, under rocks and some artifacts are also used as resting-places. The breeding den is dry and lined with reeds, grass, moss and so on.

On land Otters stand on their hind-legs, using the tail as a third leg or strut, to peer over obstructions. When bounding over the ground the back is arched. The track is usually well defined, showing the web and all 5 toes

(92a). Considerable distances are travelled over land in winter, searching for open water. Also the hunting territory of each animal extends over a wide area and several miles may be covered in a night as the animal moves from one part of its range to another. In the water Otters move with great facility: the body and tail are very flexible, and all four legs are used as paddles. When living close to large lakes hunting is almost exclusively along the banks and not out in open water.

Breeding is not restricted to any particular season; the gestation period is approximately 9 weeks and newborn young can be found from January until far on into the summer. The young remain in the den for about 8 weeks, after which they follow the mother around for several months. There is one litter in the year.

Otters feed principally on fish, and to a small extent only on crustaceans, frogs, aquatic rodents and birds. They may do considerable damage to fishing interests, particularly when they get caught in nets and traps; they are sometimes found drowned in the latter. The Otter has been persecuted in many areas both for the value of its fur and as a predator on fish. It is also hunted for sport.

93 Badger
Meles meles

Identification: Body-length 60–90 cm (23–35 in.); tail 11–20 cm ($4\frac{1}{4}$–$7\frac{3}{4}$ in.); shoulder-height about 30 cm (11 in.); weight usually 10–15 kg (22–33 lb), but may be considerably more in autumn. Looks mainly grey, but darker on the underside, and has a white head with a broad black stripe through each eye. Eyes small. Ears short, with white tips. Legs short, powerful and well-adapted for digging. Each foot has 5 toes with long claws (93b).

Distribution: Throughout most of Europe—including the British Isles—extending eastwards through northern Asia to Japan. Absent from certain islands in the Mediterranean and from the northern and western parts of Scandinavia.

Habitat: Typically where woodland or scrub adjoin cultivated land. The burrows, known as sets, are excavated in various kinds of soil—sandy soil being the easiest to dig—but low-lying, damp areas are avoided.

Habits: The set is normally a complex of tunnels and chambers, often with several entrances, excavated by the Badgers. Other animals, including the Fox, may share the same complex of burrows, but in this case each species

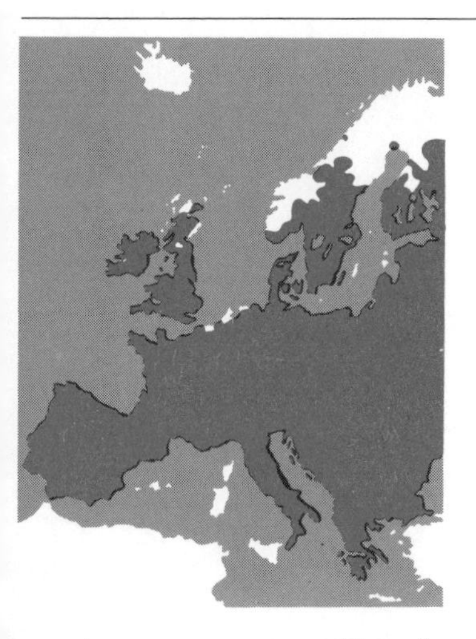

occupies a separate area. There is nearly always a large mound of excavated earth outside the main entrance to the set and there is often a distinct path running from the mouth of the set out over the mound. By contrast, the entrance to a Fox's earth lacks this path and the immediate vicinity is usually littered with the remains of prey. Badgers' sets frequently have a number of chambers. The main one, which could be described as the living quarters, is lined with dry grass, moss, bracken, leaves and so on; this bedding is changed frequently. Smaller chambers may be used as breeding dens or sleeping quarters. In large sets small chambers are used occasionally as latrines, but usually the faeces are deposited in shallow pits above ground not far from the set. The appearance of the droppings varies considerably, depending on the type of food.

The Badger is primarily nocturnal, although it will emerge for short periods in daylight in secluded places. Usually it leaves the set at dusk and returns around sunrise. Before emerging the animal will stand at the mouth of the set for some time, taking in scent and listening carefully before coming out into the open. Distinct paths radiate from the entrance to the feeding areas. When searching for food a Badger moves slowly through the terrain and snuffling sounds can be heard as it investigates everything along the path. It has poor vision, but the senses of smell and hearing are good. Scent glands are present in the anal region. The setting of scent results in a musky odour and undoubtedly has significance in the marking of territory.

Diet is very varied, indeed the Badger is often described as omnivorous. Earthworms form a large part of their diet in spring; in summer many kinds of insects, snails, frogs, small mammals and their young are eaten. Nests of wasps and bees are dug out, and the grubs are eaten with apparent relish. Large amounts of plant food are also eaten. Sometimes oats are a favoured item of diet and damage is caused by the animals rolling and playing in standing crops. Planted acorns may also be dug up and eaten. Now and again an incubating bird and its eggs may be taken and carrion is also eaten. By the autumn the Badger has laid down a thick layer of fat, which helps to tide it over the winter when the rate of activity is reduced. Although much of the time is spent in the set, this is not hibernation in the true sense of the term and even in the most severe winters Badgers will make short excursions above ground.

The breeding habits are unusual in

that although mating may take place at any time from February to October, this does not always result in fertilization of the eggs. Fertilization usually takes place in the period February–May, but embryonic development, which takes 7–8 weeks, does not start until December–January, thus there is a delay of up to 9 months. The young are born in early spring and do not usually come above ground until May. There may be 1–5 young in a litter, but the average is 3. The young are suckled for about 3 months and at the weaning stage they are fed for a time on regurgitated food. Having accompanied the mother on foraging excursions, they start finding their own food but remain with the mother at least until the autumn and sometimes even through the winter. Sexual maturity is achieved at an age of 1½–2 years.

Badgers have an extensive repertoire of sounds, including whickering, purring, various growls and squeals. There is also a frightening yell or prolonged scream, the significance of which is not yet understood.

94 Wolverine or Glutton
Gulo gulo

Identification: Heavily built, 'shaggy' appearance due to dense fur with long guard hairs. Body-length 70–83 cm (27–33 in.); tail 16–25 cm (6¼–9½ in.); shoulder-height about 40 cm (15 in.); weight usually 10–14 kg (22–30 lb) but may reach 30 kg (65 lb). Colour dark brown but a paler stripe runs along the flanks to the root of the tail and there is a greyish-white area on the forehead.

Distribution: In the mountain ranges of Scandinavia, extending northwards

to the Arctic Ocean. Also in northern and north-eastern Finland and eastwards through the most northerly parts of the Soviet Union and Siberia to North America.

Habitat and Habits: Mainly nocturnal, but also active during daylight. Moves at a rather clumsy gallop, has great stamina and can cover several miles during a night's hunting. Also climbs and swims well. The tracks are large and broad, with a distance of 40–60 cm (15–24 in.) between each group. The den may be in a natural cavity under a fallen tree, in a rock crevice or at the end of a tunnel dug in the snow. Outside the mating season Wolverines live solitarily and occupy a large home range—up to 1,800 sq km (680 sq miles). Mating may take place from about April to July. As in many other mustelids there is a long period of gestation, probably due to delayed implantation, and the 2–3, or sometimes 4, young are born in February–

April of the following year. They show interest in all carrion. Other food consists of small rodents (particularly lemmings) and hares, birds, insects and berries. Reindeer and other deer may also be killed and eaten in winter. Caches of food are sometimes placed in a tree.

Dogs

The carnivores belonging to the dog family or Canidae usually have long, slender limbs, small feet and a bushy tail. The claws cannot be retracted. The dentition is almost complete, with a total of 42 teeth, see fig. 98a. There are usually 2 or 3 well-developed molars behind the carnassial in each half-jaw. These cheek teeth have rough biting surfaces and are used for crushing hard food, including bones. The family has 5 species in Europe.

95 Wolf
Canis lupus

Identification: Body-length 110–140 cm (43–54 in.); tail 35–50 cm (12–19 in.); shoulder-height 70–80 cm (27–31 in.); weight 20–25 kg (44–55 lb). Coat colour varies from whitish-grey to yellow-grey with a reddish tinge.

Distribution: Now present only in northern, eastern and southern Europe, although formerly also in western and central Europe. Populations continue to survive in Portugal, Spain, France, Italy, Poland, the Balkans and in northern Scandinavia, but in many of these areas numbers are decreasing. Wolves are still common in many places in the Soviet Union and Siberia, and also in North America. Individuals are found wandering several hundred miles from the breeding areas. A few of these, for instance, are reported from Germany.

Habitat: Originally in widely varied terrain, chiefly in open country and woodland, but now confined to more remote areas where there is less chance of persecution by man. The presence of prey in adequate numbers is now more important than the nature of the terrain. Found in the lowlands in forests, marshland and steppe country and also in the mountains up to the alpine zone.

Habits: Sedentary during spring and summer; may travel long distances in winter when hunting in packs. Breeding season varies slightly according to latitude but mating in some areas takes place in February–March. Pairs usually remain together for several years or even for life. After a gestation

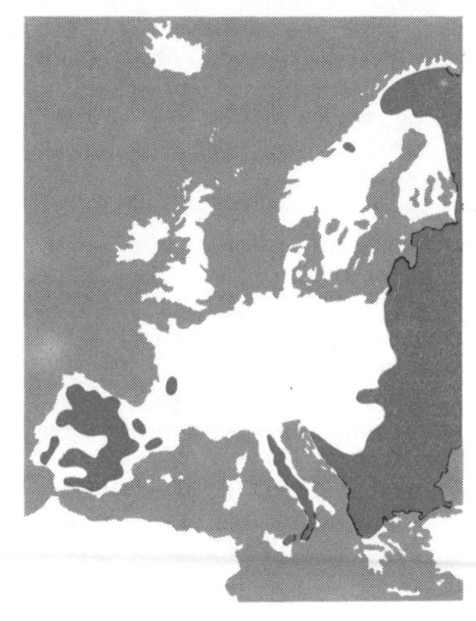

period of about 2 months the female gives birth to cubs in April–May; litter size 3–14, usually 4–6. The young are born in a well-sheltered den: a hole in the ground, a crevice among rocks or under the roots of a fallen tree, sometimes even in a haystack. The cubs are blind until they are 12–13 days old; they are suckled for 4–6 weeks but may start leaving the den at 3 weeks old. At first the female remains with the cubs while the male brings food to her; later, both parents share in getting food for the cubs. The latter start to join in the hunt when they are about 6 months old. In the late autumn a wolf family may consist of the two parents, the litter of the year and young from the previous year. Wolves become sexually mature at an age of 2–3 years.

They now hunt principally at night and catch their prey either by taking it by surprise or by a rapid but brief chase; several individuals may co-operate. Distances of more than 50 km (30 miles) are often covered during a night's hunting. Wolves usually move at a steady trot and the tracks are in a straight line, with a distance of 50–70 cm (19–27 in.) between each group. Large and small prey are attacked. Members of the deer family are taken, particularly reindeer and elk in the northern regions; various small mammals and birds are included in their diet as well as domestic animals, such as cattle, sheep, goats and dogs. Carrion is also eaten. When food is scarce extensive migrations take place. The Wolf has a repertoire of calls which resemble those of a dog, including the well-known howling, which is typically heard during October–December.

96 Asiatic jackal
Canis aureus

Identification: Smaller than the Wolf (95) and redder. Body-length 84–105 cm (33–41 in.); tail 20–24 cm (8–9½ in.), about one-third of the body-length; shoulder-height about 50 cm (19 in.); weight 10–15 kg (22–33 lb).

Distribution: A south-eastern species which in Europe is found only in the Balkans: Yugoslavia, Albania, Greece and Bulgaria, with occasional wanderers in Rumania and Hungary. Also occurs in North Africa and large areas of southern and south-western Asia.

Habitat: Mainly associated with low-lying country where there is plenty of cover, often in thickets of reeds and bushes along the banks of rivers and lakes; also in areas with ravines and caves. More tolerant of human habitations than the Wolf.

Habits: Usually lives in natural cavities in the ground, under the roots of trees or bushes. Also makes use of the burrows of other animals, such as the Fox and Badger. Well-defined paths lead to and from the den. Jackals feed mainly on rodents and birds, but will also take reptiles, insects, fruits and berries. Carrion and offal also form an important part of the diet, particularly in the vicinity of human habitation. Numbers of poultry are also taken.

Mating takes place in January–February, but otherwise their breeding habits are very similar to those of the Wolf. The Jackal is known for its characteristic howling which is heard particularly during the evening. Mainly active at night but also moves around in daylight in places where it is undisturbed by man.

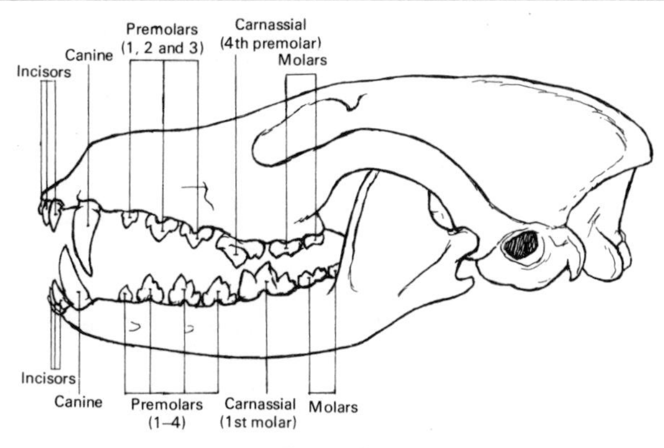

Fox skull.

97 Raccoon-dog
Nyctereutes procyonoides

Identification: Body-length 55–70 cm (21–27 in.); tail 18–25 cm (7–9 in.); weight up to 7·5 kg (16 lb). Fur dense, soft and long; various shades of black, white and grey-brown. The length of the fur on the sides of the head suggests a ruff. Ears relatively short with rounded tips.

Distribution: Originally native to eastern Asia, whence it was introduced into the Soviet Union in the 1930s as a fur animal. It has subsequently spread into Finland, Sweden, Poland and Rumania.

Habitat: Small woods and slopes with scrub vegetation close to meadows and marshland; also in reed-beds and undergrowth along the banks of rivers and lakes.

Habits: Mainly a nocturnal animal, which spends the hours of daylight in an underground den. This may be dug by the animal itself, but is often taken over from burrows made by the Fox or Badger; natural cavities among boulders or holes in trees close to the ground are also used. Hibernates from November to February–March, but this is only continuous in those individuals which have laid down sufficient fat reserves during the autumn. Hunger may force individuals into activity during the winter when the weather is favourable. Feeds on rodents, reptiles, frogs and toads, snails, insects, fish, fruits and berries; during the winter carrion and offal are sometimes included in the diet. Mating takes place after emergence from hibernation. The breeding season varies according to latitude but may start in February. The gestation period is 8 weeks and usually there are 6–8 young in a litter. Both parents help to rear the young which become independent by September–October, although some families may hibernate together.

98 Common red fox
Vulpes vulpes

Identification: Body-length 58–85 cm (23–33 in.); tail 35–55 cm (13–21 in.), very bushy and usually white at the tip; shoulder-height 35–40 cm (13–15 in.); weight 3–8 kg ($6\frac{1}{2}$–17 lb). Colour varies to some extent: on the Continent typical specimens are reddish yellow-brown on the back, white on the neck and chest, while the rest of the underside is whitish-grey. There are several variants, including a dark form which has black on the underparts and lower limbs. Head triangular, ears and muzzle pointed. The impression of the foot (98b) is longer and narrower than that of a small dog, (98c). The undersides of the feet are not furred and the impressions made by the claws are usually well defined.

Distribution: Throughout Europe—including the British Isles—also in North Africa, large parts of Asia and North America.

Habitat: Mainly associated with woodland but also in open country and has adapted well to living in built-up areas in parts of its wide range.

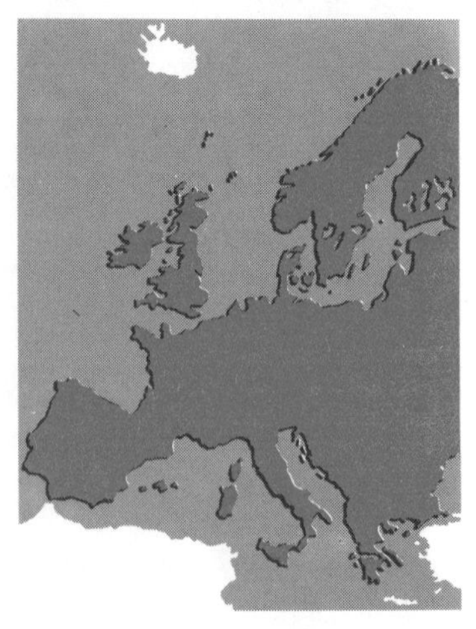

Habits: Den or 'earth' usually has several entrance-holes. The burrows may be dug by the Fox itself, but often a deserted Badger set is used. A large burrow complex is sometimes occupied by both species but in this event each has separate living quarters. There are no well-defined paths radiating from a Fox's den and there is no well-worn track to the entrance-holes as in a Badger's set. When cubs are present below ground, there may be numerous remains of prey scattered around outside, including such items as feathers, bones and household scraps. Very young cubs may be moved from one den to another if there are signs of disturbance by man. When older they are often moved to a burrow with one entrance; this summer den may be in a field sown with corn. Foxes are mainly active in the hours around sunset during the summer but in winter activity may be spread over the whole 24 hours, depending on proximity to human habitations; they are usually active for about 7 hours, but during the breeding season this may continue for up to 12 hours. The male or dog Fox normally leads a solitary life once the breeding season is finished and the young are able to catch their own prey. When on the hunt for food the animal generally moves along at a steady trot, stopping frequently for reconnaissance and relying largely on its excellent sense of hearing. The tracks lie in a straight line with a distance of 20–35 cm (8–13 in.) between the individual groups. Foxes can also run fast for considerable distances and will also swim readily. They often sleep out in the open by day, resting in some sheltered spot, particularly if their fur has become wet. Food consists mainly of small rodents, and to a lesser extent of birds, their eggs and young, and insects. During the summer large numbers of beetles are eaten, especially dung-beetles and the Fox's droppings are then coloured bluish-violet. Fruit and berries form part of their autumn diet. Rubbish tips are visited regularly and in built-up areas dustbins are even raided. When rearing young the female or vixen may make daring raids on poultry in full daylight. Foxes also visit the shore-line and in severe winters they will venture far out on the ice in search of dead waterfowl. Sick and wounded animals as well as carrion are taken. Food is sometimes buried for use in the winter.

Mating takes place in late winter, usually January–February; the gestation period is about 52 days and there are 3–12, usually 3–5 young in a litter. The cubs have dark brown fur and live in or near the den for about 8 weeks, while the vixen, assisted by the dog brings food to them. Later on the cubs accompany the vixen on hunting trips, until by summertime the family breaks up; the young are sexually mature in their first winter.

Foxes have scent glands in the region of the tail and anus, and the smell of fox can be detected by the human nose. There are several characteristic sounds including a bark—resembling that of a dog but higher and sharper—and a scream associated with the mating period.

99 Arctic fox
Alopex lagopus

Identification: Body-length 50–65 cm (20–25½ in.); tail 28–33 cm (11–13 in.), very bushy with a tip that is not white in summer; shoulder-height about 30 cm (11 in.) weight 2·5–6 kg (5½–13 lb). In summer greyish-brown or greyish-yellow on the back, paler on the underparts (99a); in winter either white or cream-coloured all over (99). A colour variant, known as the Blue fox (99b), remains a uniform blue-grey throughout the year. Ears very short and rounded. Feet densely furred on the underside.

Distribution: An arctic species which in Europe occurs only in the most northerly parts of Scandinavia and southwards along part of the mountain chain. Extends eastwards through the northern Soviet Union and Siberia to North America. Also occurs in Greenland, Iceland, Spitsbergen and other islands in the Arctic Ocean.

Habitat: Arctic tundra, and high up in the mountains above the tree limit. Lives in bare, rocky terrain and on slopes along rivers and lakes where the soil is suitable for digging.

Habits: The burrow system of the Arctic fox may be very extensive, with 50–100 or more entrance-holes and covering an area of up to 200 sq. m (140 sq. yards). Several families may live in such a complex. Natural cavities among rocks are also used as dens. The entrance-holes measure about 20 × 30 cm (8 × 12 in.) and the approximate dimensions of the main chamber are 1 m (3 ft) in diameter and 20 cm (8 in.) high. The individual burrow systems are often a considerable distance apart—5–10 km (3–6 miles) or more. The Arctic fox is omnivorous, but the principal food items are rodents, especially lemmings. They also take birds, their eggs and young, and fish, carrion washed up on the shore, also any offal or scraps; berries are also part of their diet.

Mating takes place in February–April and the young are born about 52 days later. The number of cubs varies considerably and is influenced to some extent by the availability of prey species for food. For instance, in a good lemming year there may be 7–10 or more cubs in an Arctic fox litter; in poor years far fewer, and in very bad years many of the females produce no young. The family remains together until the autumn when the members split up. As fecundity is dependent upon food supply the numbers of Arctic foxes fluctuate considerably from year to year. During the light arctic summers the foxes are active, with short intervals of rest, throughout the 24 hours, but in autumn and winter activity takes place only at night. Vocal sounds are similar to those of the preceding species. The Arctic fox is considerably less timid than the Red fox and may even approach and follow hunters and trappers at close quarters. The pelts are highly valued in certain areas.

Bears

Members of the bear family or Ursidae are large, heavily built carnivores, usually with uniformly coloured fur. They walk on the whole surface of the foot, and have 5 toes on each foot. The claws cannot be retracted.

100 Brown bear
Ursus arctos

Identification: Body-length 170–250 cm (66–98 in.); tail 6–14 cm ($2\frac{1}{4}$–$5\frac{1}{2}$ in.); shoulder-height 90–125 cm (35–48 in.); weight 150–250 kg (330–550 lb), but may exceed 400 kg (880 lb). Fur colour varies from dark brown or almost black to pale yellowish-brown. Young animals often have a pale area on the sides of the neck. Head broad, with a short muzzle and relatively small ears. The tracks (100a) resemble those of a short, broad-footed man.

Distribution: Formerly widely distributed in Europe, but now restricted to certain areas of mountain regions. There are small populations in Spain, France, Italy, Austria, Czechoslovakia, Poland and the Balkans. Also still found in Norway, Sweden and Finland, where the population is now only a few hundred. Brown bears are also found in the northern parts of the Soviet Union and in North America.

Habitat: Originally associated with large, continuous blocks of conifer and deciduous forest, formerly also in woodland in the lowlands.

Habits: A shy and cautious animal which is normally solitary. Each

animal has a very large home range and it wanders extensively through this at dusk or after dark, but in remote areas there is a certain amount of activity in daylight.

Bears normally amble along, but will break into a gallop if alarmed. They can also jump, climb and swim. The footprint is large and broad.

In the autumn a den is prepared as a resting-place: the hole may be excavated by the bear but sometimes a natural cavity is used. Moss and twigs are taken in for bedding. Some individuals remain comparatively active but others retire for the winter, the body temperature is slightly reduced and no food is eaten. If disturbed when sleeping, the bear can become active immediately.

Mating takes place about May–June. The gestation period is 6–7 months, and there is some evidence of delayed implantation. The young are born in the winter den; litter size 1–3. At birth their eyes are closed and they are about the size of a rat: the cubs are suckled until the spring when they leave the den in company with the mother. Sexual maturity and full independence are achieved at an age of about 2 years.

Bears have a very varied diet, although they are chiefly vegetarian. Insects including ants, honey, beeswax, small rodents and fish, form part of their diet. Some individuals will attack reindeer, elk and domestic animals, including cattle, horses and sheep. Bears make a number of different sounds. For example, when angry or alarmed they grunt and howl. Bears may live for up to 30 years.

101 Polar bear
Thalarctos maritimus

Identification: Body-length 160–250 cm (62–98 in.); tail only 8–10 cm (3–4 in.); shoulder-height 120–140 cm (47–55 in.); weight up to 600 kg (1,320 lb), but normally 250–400 kg (550–880 lb). Fur white or yellowish-white.

Distribution and habitat: An arctic species associated with the zone of drift ice in the Arctic Ocean and along the coasts of the surrounding mainland and islands. Very rarely lone individuals will wander into northern Norway or along the north coast of the Soviet Union.

Habits: Closely tied to the sea. Lives on icebergs and drift ice or along the coast, but seldom goes farther than a few miles from the sea. Undertakes extensive migrations, partly as a passenger on drift ice and partly by its own activity. Feeds primarily on marine animals, particularly seals. On land it also eats small rodents, hares, foxes, birds and vegetable food, as well as carrion washed up on the shore. Although a powerful swimmer that can remain submerged for up to 2 minutes the Polar bear is not particularly skilled at catching its prey in the water, as it moves rather slowly in this element. Seals are therefore caught on the ice or at their breathing-holes. Prey is killed by a blow from the front paw. The soles of the feet are densely furred and on firm ground it normally ambles along but is also capable of running fast for short distances.

Mating takes place in spring or early summer, but development does not begin until the autumn, a month before the females retire to a hole in the snow for the winter. The young are born in this winter den in December–February. At birth the cubs are the size of a rat: they are completely helpless, their eyes are closed and their fur only poorly developed. It is not until they are 1½–2 months old that they become more active, even leaving the den but remaining in the vicinity. The female usually starts bringing them more solid food when she breaks out from the den. By April the cubs accompany the mother on short excursions. The cubs continue to be suckled until they are about 1 year old, but remain with the mother until their second year. The females are sexually mature in their second or third year, males in their fourth year. Polar bears may live for up to 30 years.

Raccoons

102 Raccoon

Procyon lotor

Identification: Body-length 48–70 cm (19–27 in.); tail 20–26 cm (8–10 in.); shoulder-height 23–30 cm (9–11 in.); weight 7–8 kg (15–17 lb). Fur brownish-grey, paler on the underparts. Black facial stripes on a pale background suggest a mask. Tail black with pale rings, long and bushy.

Distribution: A native of North America which has been introduced into various countries where it has escaped from captivity. Feral populations now live in Germany, parts of the Soviet Union and it has also been reported in the Netherlands and Luxemburg.

Habitat: Forests in the vicinity of water.

Habits: Essentially a nocturnal animal, resting during daylight in a hollow tree, often at a considerably reduced temperature; periods of inactivity in winter vary from place to place and cannot be described as true hibernation. Two to six young are born in March–May after a gestation period of about 2 months. Both parents help to rear the young which become sexually mature at an age of 1–2 years. Raccoons are omnivorous but a considerable proportion of their food consists of frogs, fish, crustaceans and bivalves taken from shallow water in rivers and lakes. Reptiles, insects, birds' eggs and some plant food are also eaten. This species is valued as a fur animal.

Seals

The order Pinnipedia (seals, sea-lions and walrus) contains some 31 species distributed throughout the world. The majority live in the cold and temperate seas of the northern and southern hemisphere, and some are found in fresh waters, including certain inland lakes. Seals are highly specialized for aquatic life. The body is spindle-shaped and the greater part of each limb lies hidden in the body, so that only the fore- and hind-feet are free. The relatively long digits are webbed and the hind-feet in particular look like fins. The tail is very short. The external ears are much reduced or absent. The nostrils and ear-holes can be completely closed when diving. The eyes are usually large and positioned high up on the head, facing more or less forwards. There are 26–36 teeth, except in the Walrus which has fewer. Each half-jaw usually has 5 cheek teeth, each with one or more cusps, and they are more or less uniform in appearance. Seals live entirely on animal food which they catch in the water. In general, they only come on land when breeding and moulting; some species also haul-out on exposed land for brief but regular periods of rest. A thick layer of blubber beneath the outer skin insulates the internal organs from the cold water. The pups are particularly large and well-developed at birth. Seal's milk contains 40–45 per cent fat. Most species are markedly social and many undertake extensive seasonal migrations.

The seals are represented around the coasts of Europe by 2 families: the true seals or Phocidae and the walrus or Odobenidae.

True seals

In the Phocidae the hind-limbs are always turned backwards and are not used for propulsion on land. During swimming they are held with the soles of the feet facing towards each other, thus forming a kind of 'tail-fin'. The lateral movements of this 'fin', together with the rhythmic undulations of the body, drive the animal through the water at a considerable speed. The fore-limbs are used for steering and balance. Both fore- and hind-feet have claws, those on the fore-feet being long, powerful and slightly curved. On land the true seals can only drag themselves along on the belly with considerable effort. The coat is short and dense.

103 Common seal
Phoca vitulina

Identification: Body-length 145–195 cm (56–76 in.); tail 7–9 cm ($2\frac{1}{2}$–3 in.); weight 50–130 kg (110–285 lb). Coat silky-soft and very variable in colour. Back often yellow-grey with an olive-green tinge and numerous, irregularly arranged brownish-black spots which are all about the same size. As in the next species, the head has a relatively short muzzle which is clearly demarcated from the forehead. The cheek teeth are positioned at an angle to the longitudinal axis of the jaw (103e). The whiskers, known technically as vibrissae, are white.

Distribution: Found along the coasts of western Europe from north

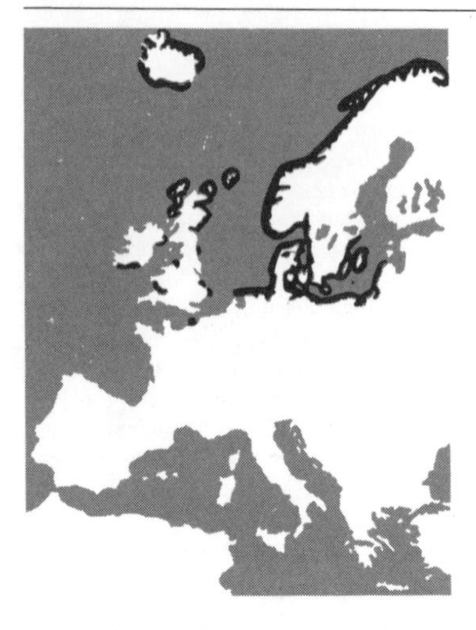

Portugal to Finmark, around the British Isles, Faeroes and Iceland and in the southern part of the Baltic Sea. There are closely related forms in the western Atlantic and in the Pacific Ocean.

Habitat: Coastal and relatively shallow waters.

Habits: Spends a large part of the day hauled-out on land, but only in places where there is very little disturbance and where it would be difficult to take the animal by surprise. During the breeding season, in June–July, this seal gathers in large numbers at a few selected sites, but at other times the colonies are smaller and more scattered. In Britain, the main breeding areas are off East Anglia, in various parts of Scotland including the Outer Isles and also off northern Ireland. Several characteristic resting positions (103c) can be observed from a distance with the aid of binoculars. The pup may be born in the sea or on a rock or bank that is only exposed for a short period. In contrast to other species, the pups of the Common seal are rarely born with the white woolly pelt, often referred to as puppy-coat; this is either shed in the uterus before birth or immediately afterwards. The pups are able to accompany their mothers in the water after only a few hours. At birth they are almost 1 m (about 3 ft) long and are suckled for about one month. Suckling takes place both on land and in the water. The young become sexually mature when they are about 4 years old. Mating has been recorded from early summer to late autumn. It takes place in the water, usually on or near the bottom at a depth of 4–8 m (12–25 ft) and is accompanied by various courtship displays, which are never seen on land. After mating there is delayed implantation, the fertilized egg not starting to develop until the following winter. Diet consists of various kinds of fish, especially flatfish and cod, as well as crustaceans; feeding is mainly during the hours of daylight. Large prey is held by the fore-feet while being eaten. The daily food requirement is about 5 kg (11 lb) of fish. This seal can reach a speed of about 35 km (21 miles) per hour and it normally remains submerged for 7–10 minutes each time it dives.

104 Ringed seal
Pusa hispida

Identification: Body-length 100–185 cm (40–72 in.); tail about 10 cm (4 in.); weight 40–125 kg (88–275 lb). Plumper than the preceding species, especially towards the rear; in winter the blubber tends to be very thick and may account for a considerable proportion of the total weight. The coat is coarse. Back grey-black or brownish-black with a conspicuous pattern of straw-coloured rings. Sides and belly whitish-yellow, sometimes with indistinct markings. Muzzle short and blunt as in the Common seal (103). The cheek teeth are smaller and are aligned parallel with the long axis of the jaw (104b). Whiskers brown.

Distribution: Mainly in the Arctic Ocean, but there are isolated populations in the inner Baltic Sea, the Gulf of Bothnia and the Gulf of Finland, and in Lake Saimaa and Lake Ladoga. Related forms in the Caspian Sea and Lake Baikal. Now and again vagrants from the Arctic Ocean are found farther south: rarely off the British Isles, occasionally along the west coasts of Norway and Sweden, more frequently on the coast of northern Norway.

Habitat: In winter in ice-covered waters, living on or under the ice. In summer on banks and rocky reefs near the coast.

Habits: When the waters freeze these seals make breathing holes by pushing the ice up to form a small air-filled dome. The breathing holes are kept open the whole time, even when the ice is over 1 m (about 3 ft) thick: the seals visit each hole several times an hour and the water that has become somewhat chilled in the meantime is displaced by the seal's body and replaced by slightly warmer sea water when it dives again. In addition, there are larger holes, which are used when the seals come out on to the ice. The pup is born during March–April, in a snow tunnel on the ice; this tunnel, which is several yards long, is made by the female in the lee of ice blocks. The pup has a thick, white woolly coat for the first 2–3 weeks and it remains in the tunnel during this time. It is suckled for about 2 months and grows to a length of 60–70 cm (24–28 in.); it then becomes independent and first starts going into the water. Breeding does not take place until the seal is several years old. Mating occurs about a month after the birth of the previous pup but true gestation does not begin until late in the summer. Diet consists of various kinds of fish and crustaceans which are hunted at depths down to several hundred metres. The Baltic population has decreased considerably in recent decades, but elsewhere the Ringed seal is still very common.

105 Grey seal
Halichoerus grypus

Identification: Total length 165–330 cm (65–130 in.); weight 100–315 kg

(220–690 lb). Mature males considerably larger than females. The colour is variable: the ground colour may be silvery-grey or dark grey, creamy yellow or fawn. Although females usually have a paler ground colour with dark spots—whereas males tend to have light spots on a dark ground—this is not a safe guide to sex identification. Adult males have thicker necks than adult females, and the male profile is convex whereas the female is straighter. Both sexes have a long muzzle which is not sharply demarcated from the forehead. The cheek teeth are conical with a single point but sometimes there are weakly developed side points (105c).

Distribution: A large population, about 35,000 individuals in Europe, lives around the British Isles, Faeroes, Iceland and western Norway, with a smaller population of about 10,000 individuals in the Baltic Sea. Other populations live on the coasts of

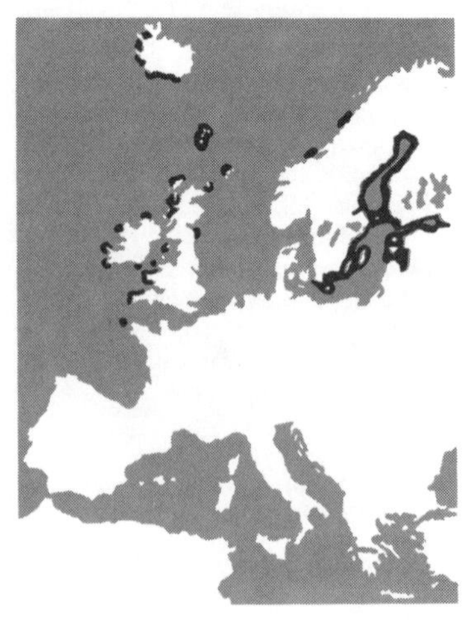

Greenland, Nova Scotia and Labrador.

Habitat: Particularly associated with rocky coasts exposed to open sea, but also found on sand banks and rocky reefs.

Habits: During the breeding and moulting seasons Grey seals live in large colonies of several thousand individuals or in smaller groups with only 10 or so individuals. After this they spread along the coasts and out to sea. They spend less time on land than the Common seal (103) and even then always in the immediate vicinity of water. The actual time of breeding varies considerably from place to place. Around the British Isles the main breeding season is August–December. The bulls arrive at the breeding beaches before the cows and take up their territories. At this time individual animals may remain on the breeding grounds for about 2 months without feeding. The bulls are very aggressive and fights are common, and only the strongest bulls succeed in maintaining a territory. The result is that there may be 20 times as many cows as bulls present on the breeding beaches. The females remain for only about 3 weeks, during which time they give birth to a pup and then mate again. The cows, as well as the bulls, go without food at this time. Normally only a single pup is born and it has a yellowish-white, woolly coat. It is suckled by the mother during the 3 weeks she is on land and is then left on its own. Before returning to sea, the female will have mated with one or more of the bulls present. Mating takes place on land as well as in the water. Most of the young are born in September–October but in the Baltic pupping—on the open ice—does not take place until February–March. The gestation period is therefore about $11\frac{1}{2}$ months, but the true embryonic development only takes place during the last $7\frac{1}{2}$ months. The deserted pups shed their woolly coat and go into the water where they gradually learn to catch food. During the transition period they live on their reserves of fat. They are sexually mature at an age of 4–5 years. The Grey seal is more vocal than most of the other seals and the sounds made by a colony can be heard at a considerable distance. They feed on fish, crustaceans and squid, and hunt both by night and day.

106 Monk seal
Monachus monachus

Identification: Total length 230–380 cm (90–150 in.); weight 300–320 kg (660–700 lb). The back and sides are a uniform dark brownish colour with a greyish olive-green tinge; belly pale. Spots are often indistinct or absent.

Distribution: Widespread in the Mediterranean, but always very local. The largest population is in the Adriatic Sea. Also found in the Black Sea and in the Atlantic Ocean as far west as the Canary Islands and Madeira.

Habitat: Sheltered places along shores and rocky coasts.

Habits: This seal appears to be somewhat lethargic and trustful. These characteristics, together with a tendency to aggregate in small and limited areas, has made it more vulnerable than other species. The total world population has been estimated as no more than 5,000. Little is known about their habits.

107 Greenland or Harp seal
Pagophilus groenlandicus

Identification: Total length 155–220 cm (61–86 in.); weight 115–180 kg (260–395 lb). The male, which is somewhat larger than the female, is yellowish-white or greyish-yellow, with a brownish-black area running from the shoulders along the sides to the tail; most of the head is also dark. The females and young have a greyer ground colour.

Distribution and habitat: Mainly an arctic species; it occurs in a belt from north-eastern Canada, through Greenland, Jan Mayen and Spitsbergen to the Kara Sea around Novaya Zemlya and is also found in the White Sea and on the coast of the Kola Peninsula. This species has three main populations: the first and largest breeds in the waters off Newfoundland, the second off Jan Mayen and the third in the White Sea. Outside the breeding season mature Harp seals undertake long migrations which take them several thousand miles from the breeding areas. A few vagrants have been recorded from time to time off the coasts of western Europe, including the British Isles; they are seen more frequently off the coast of northern Norway.

Habits: Breeding takes place during the spring—during March in the White Sea—in large, dense colonies on the drift ice and mating occurs again after the birth of the pups. There may be one or two pups in a litter, individual litters being spaced out at 5–10 m (16–30 ft) apart. The young are only suckled for about 2 weeks, after which they are left to themselves. The white coat, which makes these seal pups so attractive to fur hunters, is only in perfect condition for about 10 days; after this it becomes loose. When the moult is complete after about 14 days, the pups go into the water for the first time. Sexual maturity is achieved at 3–4 years. Diet consists of fish, particularly herring and cod, as well as crustaceans which are caught at depths down to more than 200 m (600 ft). This species also lives in large aggregations outside the breeding season; the males, females and young often keep in separate groups. During the moulting period in summer they again congregate and enormous numbers are seen together on the ice.

108 Hooded seal
Cystophora cristata

Identification: Total length 200–235 cm (80–92 in.); weight 350–400 kg (680–880 lb). Ground colour pale or dark grey. Back with irregular brownish spots. The head, neck and forelimbs are brown. In the adult males there is a pocket of skin between the nose and eyes, about 25 cm (10 in.), which can be inflated; in addition, the nasal mucus membrane can be everted from the nostril and blown up into a large, red bladder.

Distribution and habitat: Breeds in two separate regions, in the waters between Newfoundland and Greenland, and north of Jan Mayen. Outside the breeding season Hooded seals move about a great deal, either alone or in small groups, and some extend eastwards to Spitsbergen and Novaya Zemlya. They avoid the Gulf Stream as far as possible and so are only rarely seen along the coasts of western and northern Europe. They are frequently seen off the coasts of northern Norway and the Kola Peninsula. The occasional vagrant has been recorded off the British Isles.

Habits: Like the preceding species, this seal is tied to the drift ice and is also a great wanderer. One or two pups per litter are born on the ice during March–April; these weigh 11–14 kg (24–30 lb) at birth and have a white woolly pelt which is soon moulted and replaced by a delicate, greyish pelt (108b). The 'blue-back' phase of the pelt is commercially desirable and at this stage the pups are hunted annually. Mating takes place a few weeks after the birth of the preceding litter and the bulls fight for possession of the cows. At this time the expanded hood serves as a threat display. After pupping and mating, migrations take place but by the summer the seals become more sedentary and this is when they moult on the ice. Later they migrate back to the breeding areas. Hooded seals become sexually mature at an age of about 4 years.

109 Bearded seal
Erignathus barbatus

Identification: The largest seal along the coasts of Europe. Total length 220–310 cm (86–122 in.); weight 300–410 kg (660–900 lb). Almost uniformly grey-brownish, yellowish-brown or grey-blue, and palest on the belly. Whiskers particularly long and luxuriant.

Distribution and habitat: Widely distributed along the coasts of the Arctic Ocean. In Europe: only in the White Sea and on the coasts of northern Norway. Also off Jan Mayen, Spitsbergen and Greenland. In certain years Bearded seals move a long way south along the Norwegian coast. There are also a few records from the British Isles. The world population has gradually decreased: although estimates vary, there are now thought to be only 75,000–150,000 individuals.

Habits: Much less sociable than the preceding species and does not undertake long migrations. During the mating and pupping season colonies of up to 50 individuals are seen, but at other times it is usually solitary. Found mainly in the vicinity of the coast, where the water is 30–50 m (95–165 ft) deep; often swims up into the lower reaches of rivers. The young are born on the pack ice during April–May, often quite far from open water; like the Ringed seal (104), this species also maintains breathing and diving holes in the ice. It feeds mainly on bivalves, snails and crustaceans taken from the bottom, but also catches fish. Bearded seals become sexually mature at an age of 6–7 years. The hide which is very tough has been much in demand for the runners of dog sledges.

Walrus

In the walrus family, Odobenidae, the hind-limbs can be turned forwards and are used when moving on land. The coat is short and sparse.

110 Walrus

Odobenus rosmarus

Identification: Total length 300–450 cm (120–180 in.) in the male, 300 cm (120 in.) in the female; weight 700–2,200 kg (1,540–4,800 lb). Colour yellowish-brown, young animals being palest. The growth of the hair is never strong, becoming even more sparse with age. In the upper jaw the canine teeth are developed into tusks which are up to 50 cm (20 in.) long. The whiskers are luxuriant and well developed.

Distribution: Formerly numerous and widespread in the North Atlantic area, including the Barents Sea, White Sea and Kara Sea north of Europe. Excessive hunting has reduced the numbers considerably and there is now only a small population on the east coast of Novaya Zemlya and in Greenland. At intervals of several years some individuals stray southwards to the north coast of Europe and occasionally even farther south, including rare sightings off the British Isles.

Habitat: Lives on drift ice in shallow water off open coasts, and also comes on land.

Habits: Diet consists mainly of bivalves and snails which are dug up from the bottom with the help of the tusks, brushed clean with the whiskers and then crushed by the flattened cheek teeth. Walruses also hunt seals, which are seized from below at their breathing holes with the powerful forelimbs. In winter they keep a fair-sized stretch of water free of ice as the holes, used by some of the true seals, are inadequate for the Walrus. This species can dive for up to 10 minutes.

Walruses generally live in family groups consisting of an adult male, several females with young and a number of immature juveniles. The mature male acts as leader and shows considerable aggressiveness in defending the family group. Arctic hunters consider it very dangerous to wound a Walrus which may belong to a family party as the other members of the group will then attack. Mating is not restricted to any particular season in the year, as in some of the seals, and new-born pups can be found throughout the spring and summer. The young are suckled for at least $1\frac{1}{2}$ years and become sexually mature in their fifth year. Sounds made by Walruses include a double-noted call, repeated several times, which has been described as 'awook'.

Artiodactyls

The order Artiodactyla, or ungulates with paired toes, contains all the ungulates except the tapirs, rhinoceroses and horses which belong to the order Perissodactyla (ungulates with unpaired toes). A common characteristic of the 2 orders is that they lack the first toe, and that the other toes form a hoof; in addition they are specialized for a vegetarian diet. In the perissodactyls the weight of the body is carried on the central or third toe, which is much larger than the other toes, and the central axis of the foot passes through this digit. In the artiodactyls, on the other hand, the weight is carried on the third and fourth toes, and the axis of each foot runs between these two; and the second and fifth digits are poorly developed, except in the pigs. The artiodactyls are classified in two groups: the family Suidae which are less specialized and to some extent omnivorous, and the ruminants (Ruminantia) which are exclusively vegetarian and in most cases highly specialized. The European ruminants are represented by 2 families: the deer or Cervidae and the cattle, sheep and goats or Bovidae.

Pigs

In the pig family (Suidae) the metacarpals are only slightly elongated and they are not fused with one another. The second and fifth toes are relatively well developed. The number of teeth is large—up to 44. The canines, which are particularly long in the male, are like tusks and they curve outwards and backwards. The cheek teeth are very low with rough surfaces, well-adapted for chewing a wide variety of foods. The muzzle is developed into a hairless, disc-shaped snout which is very mobile and is supported internally by cartilage and a couple of small bones. The skin—except on the snout—is covered with coarse hairs and bristles; there is a well-developed layer of fat under the skin. The digestive tract is comparatively simple in structure. The number of young—and of nipples—is large; the young piglets are born with hair and can move around immediately.

111 Wild boar
Sus scrofa

Identification: Body-length 110–155 cm (43–61 in.); tail 15–20 cm (6–$7\frac{1}{2}$ in.); shoulder-height 90–100 cm (35–39 in.); weight 75–200 kg (165–440 lb). The male or boar is considerably larger and heavier than the female or sow. Head elongated, terminating in a disc-shaped snout. Compared with the domesticated pig, the body is more laterally compressed and the legs relatively longer. Coat brownish-black, coarse-haired in adults. The piglets have characteristic pale longitudinal stripes (111b). The canine teeth of both jaws are well developed, particularly in the boar; in the process of rubbing against each other the edges of the canines become razor sharp and growth is continuous throughout life.

Distribution: Formerly widely distributed throughout Europe but now only found wild in parts of central, south and eastern Europe, and eastwards into Asia. It became extinct in Britain during the seventeenth century. Fecundity is high and under abnormal conditions—for instance when no hunting takes place—the rate of reproduction is surprisingly fast. Present distribution and numbers have also been influenced by man: in some places it is valued in the interest of sport and elsewhere it is destroyed because of the damage it does to crops.

Habitat: When living wild this species is found in forest regions, particularly where these border on pasture and arable land. It also occurs in steppe country provided there are rivers with sufficient stretches of undergrowth and marsh vegetation.

Habits: The sows and young are sociable, usually living in herds of up to 50 individuals. The boar is primarily solitary and lives on its own except during the breeding season. In summer the hours of daylight are mostly spent resting in depressions in the ground, lined with dry grass and moss, preferably in a sunny place surrounded by dense woodland or scrub. In winter a number may be found sheltering in holes in the snow, packed close together. A lot of time is spent wallow-

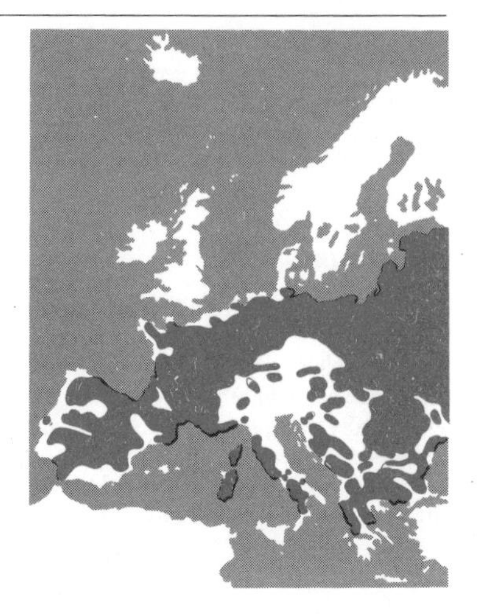

ing in the mud of small ponds and marshes, particularly in the autumn. In general they are good swimmers. They search for food in the morning and in the evening, either in the forest itself or in nearby fields, using the snout to turn over the upper layer of soil and the tusks to dig up roots. The diet consists of practically anything edible but plants form the greater part. As root vegetables—such as potatoes and turnips—are included in their diet, considerable damage to field crops may be caused in some areas. Beechmast and acorns are also eaten in large quantities. Other items include all kinds of insects, worms and snails, small rodents, nesting birds, sick or wounded mammals and carrion.

The mating season starts in late autumn and in some areas it may even continue right up to April. During this period vigorous fights take place among the older boars. The gestation period is about 16 weeks. The number of piglets in a litter is variable (2–12); the piglets are born in a well-protected depression, and can follow the mother after only a day or two. Sexual maturity is reached at an age of about $1\frac{1}{2}$–2 years and the life-span is up to about 20 years. This species leaves a strong odour but although it has a number of scent glands little is known about their function.

Ruminants

In the ruminants (Ruminantia) there are two metacarpals (the third and fourth) which are considerably elongated and fused to form the so-called cannon bone; there are also two digits forming a hoof. The limbs are generally long and slender. The teeth are highly specialized for a vegetarian diet. In the upper jaw the incisors are usually completely lacking and the canines are also often absent. In both the upper and lower jaws there is a long toothless gap between the cheek teeth and the incisors (when present). In the lower jaw the canines have moved forwards to the incisors, which they largely resemble. The cheek teeth have prominent ridges of enamel and form an extremely effective apparatus for breaking up even the toughest plant food. During chewing the lower jaw moves from side to side. The structure of the digestive tract is very complicated. After a first chewing the food—together with saliva—slides down into the first chamber of the stomach known as the rumen. The plant food undergoes a certain amount of bacterial fermentation in the rumen. From time to time some of this food passes to the reticulum, the second stomach chamber and thence up the oesophagus and into the mouth. This food, known as a cud or bolus, is chewed and then moves down again into the rumen. It may be regurgitated and chewed again. Food which has been thoroughly masticated passes from the rumen through the reticulum to the third stomach chamber, known as the omasum and finally reaches the fourth chamber or abomasum where the first stage of true digestion takes place.

The number of young is always small, usually one or two, and these are born in a well-developed state and can move about almost immediately. In Europe the artiodactyls are represented by two families: the deer or Cervidae and the cattle, goats and sheep or Bovidae.

Deer

The deer family or Cervidae contains about 50 species. Deer are characteristic of the forested areas of the whole of the northern hemisphere and also of South America. Only 5 species are native to Europe but others have been introduced and are now successfully established in the wild. With one exception the males all have antlers which are grown and shed annually and there is one species in which both sexes have antlers. In the forehead region of the skull there are 2 bony protrusions, known as the pedicles, which are covered with skin; the antlers grow from the tips of these protrusions. At first the antlers are covered with tight skin, known as 'velvet', which is richly supplied with blood. The antlers are formed of cartilage which gradually ossifies into hard bony structures. When this has taken place the 'velvet' dies off and dries out;

it is then removed by the deer rubbing or 'fraying' the antlers against young trees and bushes. The rut or onset of the breeding season starts when the antlers are fully grown. Some time later there is a breakdown of the tissue between the pedicle and the base of each antler, known as the coronet or burr, and this results in the antlers being shed. Shortly afterwards, skin grows over the fractured surface and the whole process of antler growth starts again. The second and fifth digits are much smaller than in the pigs and are positioned higher up on the leg. Deer have a number of scent glands in various places including: in front of the eyes, on the forehead, the chin and on the feet and legs. The secretions from these glands are used, among other things, for marking territory during the rutting period. There are 32–34 teeth.

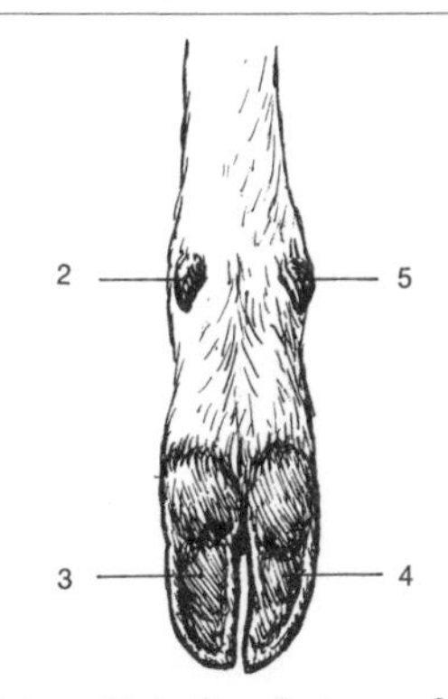

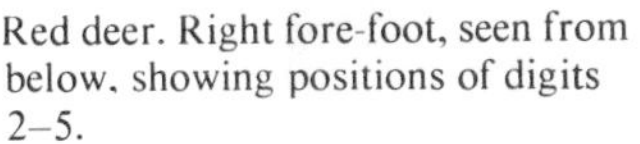
Red deer. Right fore-foot, seen from below, showing positions of digits 2–5.

112 Chinese water deer
Hydropotes inermis

Identification: Body-length 78–100 cm (30–39 in.); tail 6–8 cm ($2\frac{1}{4}$–3 in.); shoulder-height 45–55 cm (17–21 in.); weight 9–11 kg (20–24 lb). Uniform yellow-brown with dark flecks. Neither sex has antlers. In both sexes the upper canines are elongated to form tusks.

Distribution and habitat: Originally native to China and Korea, whence it was introduced into the British Isles about 1900. Latterly it became established in the wild in a few places. There are also feral populations in France. In its native range this deer is associated with swampy reed-beds but in England it lives in dry areas of woodland and open fields.

Habits: Active by day and by night, usually living solitarily. When hunted it makes characteristic leaps like a hare. The litter-size is larger than in any other species of deer and varies between 3 and 7. The young are born in May–June. Diet consists of grasses and other plants. This species has a number of scent glands, including one in front of each eye.

113 Chinese muntjac
Muntiacus reevesi

Identification: Body-length 80–100 cm (31–39 in.); tail about 17 cm ($6\frac{1}{2}$ in.); shoulder-height about 45 cm (17 in.); weight 18–22 kg (40–48 lb) in the male, about 13 kg (28 lb) in the female. Coat short and soft. Coloration very variable, from pale brownish to red-brown, richer in summer than in winter. The male has short antlers (113a) borne on high, hairy pedicles. In the male the upper canine teeth are usually slightly elongated and directed outwards.

Distribution and habitat: Originally native to eastern and southern Asia, whence it has been introduced into the British Isles, where populations have now become established in many places. Feral populations are also found in France. Lives mainly in woodland with dense undergrowth but also enters fields and gardens in search of food.

Habits: Diurnal but main feeding activity takes place after dark. Feeds on grasses, leaves, shoots and buds, including cultivated plants. This species moves about singly or in pairs. A loud, hoarse bark is heard at all times of the year, usually after dark, and is repeated at intervals. It is possible that the main rut takes place during winter but fawns have been observed in various months of the year. There are 1–2 young in a litter. This species has scent glands on the forehead between the pedicles, in front of the eyes, under the chin and in the region of the hind-feet.

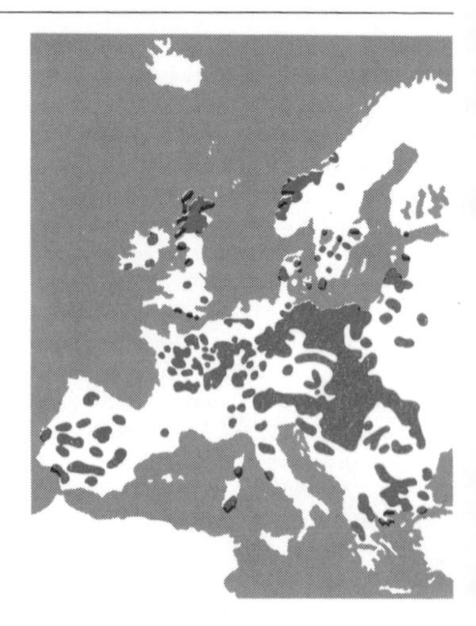

114–115 Red deer
Cervus elaphus

Identification: Body-length 165–250 cm (65–97 in.); tail 12–15 cm (4–5 in.); shoulder-height 120–150 cm (46–58 in.); weight varies according to feeding conditions, the male or stag 100–255 kg (220–560 lb), the female or hind somewhat less. Coat red-brown in summer, more grey-brown in winter. Rump patch and underside of tail cream-coloured (115b) rather than white as in next species (116b). The calf (115a) usually has white spots on the body. The stag's first antlers are formed at the beginning of its second year of life; they are short and unbranched (114b). The number of branches or points on each antler increases gradually year by year, until these are 5–8 (114c–f), sometimes more. The stag is named according to the number of its points: thus, a stag with 12 points (6 on each antler) is known as a royal. The number of points is not associated only with age but depends also upon nutritional conditions. Full development of the antlers is not achieved until the stag is 9–10 years old. The antlers of older animals are shed every year early in spring but the first antlers of young stags are cast a little later. The new antlers, covered with hairy skin or 'velvet', grow during the course of the summer (114a). In July–August the adult stag rubs off the 'velvet', which by then has become dried, by 'fraying' against trees and bushes.

Distribution: Found over large areas of Europe and eastwards through Asia to North America, where it is known as the Elk or Wapiti. In the British Isles, Red deer are particularly abundant in parts of Scotland.

Habitat: Normally associated with forest, both coniferous and deciduous, with access to open terrain. In Scotland and Norway, however, they live high up on mountain moorland for much of the year.

Habits: During the day Red deer tend to lie up in the forest or up on the hillside, keeping in the shade during the summer and seeking shelter in the winter. In the evening they leave their resting-places and follow well-established paths to the feeding-grounds and drinking-places. They also wallow in muddy pools or peat hags. When on the move their usual speed is a steady trot but sometimes they break into a gallop. They jump well and are strong swimmers. For a large part of the year the hinds, young stags and calves live in large herds, led by an old hind. The mature stags either live on their own or form small herds. In May–June the pregnant hinds leave the herd and find a sheltered spot where they give birth to their calves. After a couple of weeks the hinds and the new-born calves form into herds again and are joined by the hinds with calves of the preceding year. In September–October, as the rutting period approaches, the mature stags move to the rutting areas. Individuals often establish the same mating territory year after year. Wallowing activity increases and scent from glands in the region of the eyes is used for marking territory. It is at this time that the characteristic 'roar' of the stag is heard. The hinds are rounded up and having gathered as many as can be held together, the stag attempts to

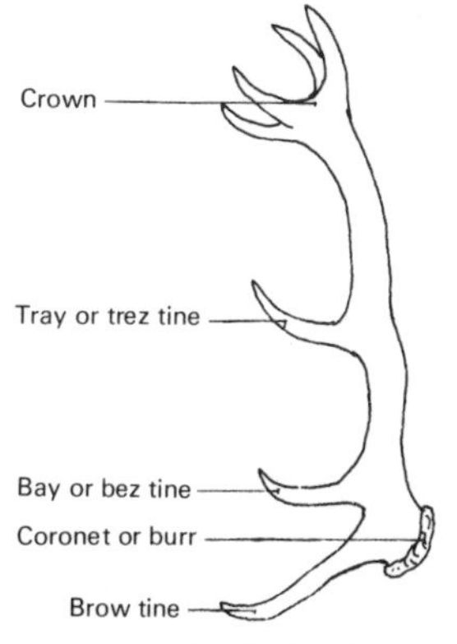

Red deer stag. Left antler.

drive off all male intruders. Young and weaker stags withdraw when threatened but fights occur when the rival stags are more evenly matched. Such fights usually end as the weaker animal loses strength and withdraws to a distance, leaving the victorious stag with the hinds. Towards the end of the rutting period when the successful stags have lost some of their energy the young stags may get an opportunity to mate with those hinds which come on to heat late. Each individual hind is receptive for only a day or two. The hinds become sexually mature in their third year. The gestation period is 230–240 days. Each hind normally produces only one calf a year. The calf has a high-pitched bleat which is answered by the nasal bleat of the mother, a sound which resembles that of a sheep. Red deer may live to an age of about 30 years.

Diet consists of grass, heather and other moorland plants, various fruits, fungi and lichens as well as the leaves and shoots of bushes and trees. As this species also eats bark and uses trees for 'fraying', considerable damage may be caused in young forestry plantations. Field crops may also be affected as corn, turnips and potatoes are included in their diet.

116 Sika deer
Cervus nippon

Identification: Body-length 110–130 cm (42–51 in.); tail 10–15 cm (4–5 in.); shoulder-height 80–90 cm (31–35 in.); weight 40–45 kg (88–100 lb). Coat red-brown in summer with rather faint spots on the flanks; in winter, usually uniform dark brown. Rump patch white—not creamy—outlined by dark hair across the top and down the sides, emphasizing the heart-shape (116b). The tail is white, with a narrow dark stripe on the upperside which is only faintly marked by comparison with the Fallow deer. The antlers of the male (116a) resemble those of the Red deer stag (114–115) but tend to be simpler, with a maximum of 8 points.

Distribution and habitat: Originally native to eastern Asia but about 100 years ago several forms of this deer were introduced into central and western Europe from Japan and northern China. Now feral in several countries. The species is associated with deciduous woodland, where there are thickets and access to clearings, and with unthinned plantations of young conifers. It also occurs in marshy areas where there is sufficient cover.

Habits: Not as gregarious as the Red deer but otherwise very similar. During the rutting period, October–November, the males give a characteristic whistle which rises and falls and ends in a grunt. This call is repeated 3 or 4 times and is then followed by a long pause. Breeding is much the same as in the Red deer and a single calf is usually born May–June. Diet is similar to that of the Red deer, consequently forestry and farming interests may be adversely affected. Sika deer are particularly shy and retiring in their behaviour and once they have become established in an area it is almost impossible to get rid of them. This species has scent glands in the region of the eye and on the lower part of the leg.

117 Fallow deer
Dama dama

Identification: Body-length 130–160 cm (41–62 in.); tail relatively long, 16–19 cm (6–7 in.) and with a very dark stripe on the upper surface, white on the underside; shoulder-height 80–100 cm (31–39 in.); weight 60–85 kg (132–185 lb) in the male or buck, 30–

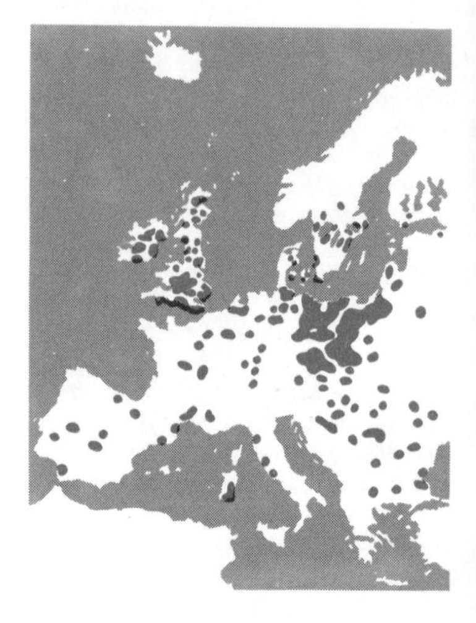

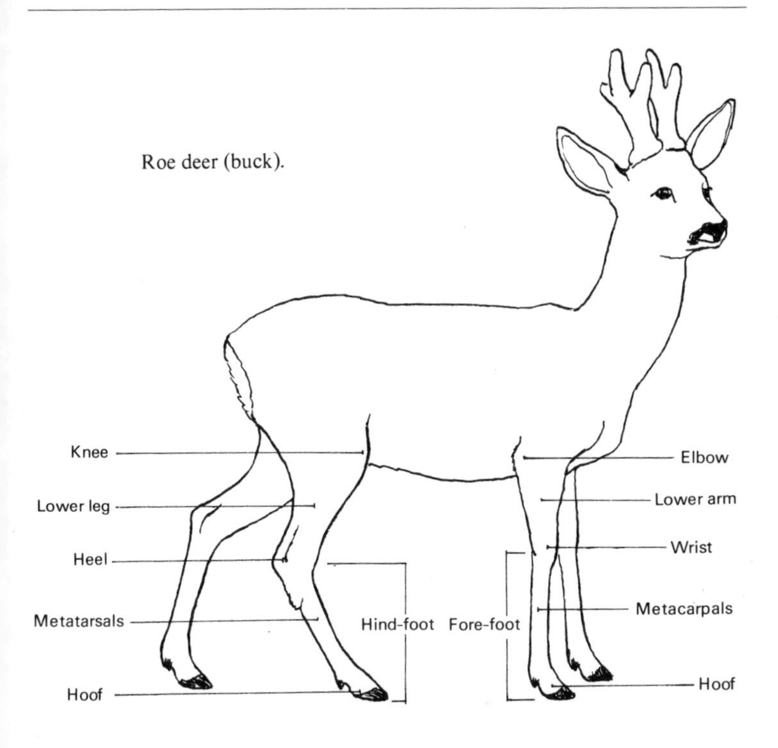

Roe deer (buck).

50 kg (66–110 lb) in the female or doe. Ground colour of the summer coat varies from fawn to red-brown, with a dark stripe along the back and numerous small white spots on the back and flanks (117a); the winter coat looks much more drab and there is only a faint indication of spots. Rump patch pale and edged with black (117c). There are several colour variants, including some very dark (117b) and almost white forms. A fully grown buck, about 6 years of age, has shovel-shaped antlers; these are shed annually in April–May. The first antlers, short and unbranched, develop when the buck is nearly 2 years old; these are shed in June. The new antlers are fully formed in August and by September they are already clean and free of 'velvet'.

Distribution: This species is known to have been well established in various parts of Europe, including Britain, for centuries. It is usually regarded as native only to Asia Minor although some authorities consider it may also have been native to the Mediterranean area of Europe. Today there are populations spread over most of Europe, including Britain and Ireland,

and as far north as central Sweden and southern Finland.

Habitat: Now occurs both in the wild and in deer parks. Often in wooded country, in the vicinity of parks from which they have escaped, spreading into mature stands of deciduous trees or in mixed woodland, usually with some intermediate layer of cover and with access to grassy glades or fields.

Habits: Although often seen grazing in parks during the day, the Fallow deer is rather shy and retiring. Main feeding activity usually starts in the evening and apparently ends soon after dawn. Herds often move in single file and when trotting or galloping, they look somewhat ungainly but they can jump and swim well. As in the Red deer (114–115) the does and young animals live in herds for much of the year while the older bucks usually live apart, either alone or in small parties. The main rutting period is in October–November. Bucks challenge each other with rhythmic snorts and grunts and individual territories are taken up in much the same way as in Red deer. Scent from a gland on the forehead, and also from urine, plays an important part in marking territory. The forehead and the antlers are rubbed against trees, and also on the ground where earth has been scraped and used as a place for urinating. Thus the scent of each buck is effectively distributed. Wind helps to disperse the scent and the weaker bucks establish territories downwind of the stronger bucks. The gestation period is about 240 days. Usually a single fawn, heavily spotted, is born in May–June. Diet consists mainly of grasses, herbaceous plants, seeds and berries. Fallow deer also eat field crops, as well as foliage and bark of trees and bushes. Both the buck and the doe will give a hoarse bark when alarmed. Bleating calls are also heard from does and fawns.

118 Roe deer
Capreolus capreolus

Identification: Body-length 95–135 cm (37–53 in.); external part of tail very short, only 2–3 cm (about 1 in.) and indistinguishable from rump patch; shoulder-height 65–75 cm (25–30 in.); weight 15–30 kg (33–65 lb). Coat red-brown in summer (118a), grey-brown in winter. Melanistic and albinistic forms known to occur but they are rare. Rump patch is noticeably white in winter and when excited the tufted hairs are raised and spread out (118c). The antlers of the buck are vertical, short and comparatively simple with 2, 4 or 6, exceptionally 8 points. The development of the antlers increases with age but is also

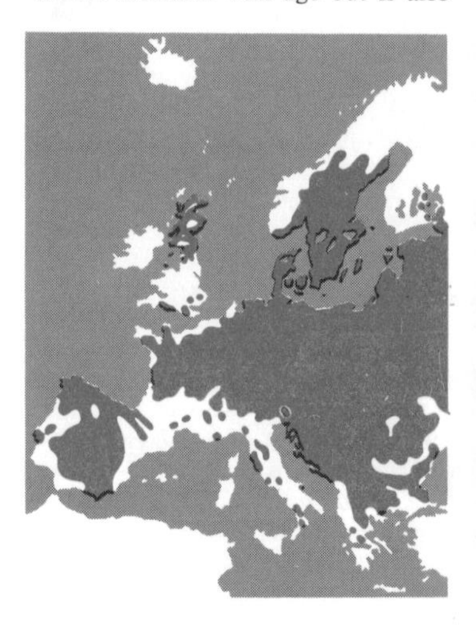

associated with nutritional conditions. The antlers are shed every year in November–December. The new antlers are fully developed in March–April and are rubbed clean in April–May; antlers of first-year bucks are cleaned slightly later. The kid (118b) is red-brown, paler on the underside, and spotted with white on the flanks.

Distribution: Found throughout most of Europe, but absent from parts of southern Europe and northern Scandinavia. Native to Britain but absent from Ireland.

Habitat: Associated mainly with immature woodland or forest with plenty of undergrowth but also spreads out into cornfields, dry reed-beds, scrub and similar places with adequate cover.

Habits: Mainly active in the morning and evening and daylight is usually spent lying up in cover. In secluded places, however, they also feed in the middle of the day. Remarkably quiet and soft-footed unless disturbed. Take flight in a series of startled leaps, bounding away but often stopping abruptly to look back. Swims well. The bucks are territorial for much of the year. The marking of territory includes rubbing against trees and bushes and scraping up earth. Scent glands are present in this species on the forehead and in the region of the feet and legs. The rutting period is in July–August. Shortly before the does come on heat there are fast chases in which the doe gallops away with the buck in hot pursuit. The buck rounds up the doe, driving her in a circle or a figure of eight, thus making well-trodden rings, about 1–3 m (3 ft) in diameter, which often have a bush or similar obstacle in the centre. Presumably this is part of a courtship display. Mating takes place chiefly in August but at this time of year true development of the embryo is delayed for some months. The gestation period is about 10 months. The kids, usually 2, are born in May–June. For their first 2 weeks they are left on their own for much of the time, lying up in cover while the mother feeds. Later they accompany the mother until the following spring. In winter, Roe deer are often found in small family groups, but in general this species is not gregarious.

Diet consists chiefly of leaves and shoots of various trees and bushes, consequently they can do considerable damage in forestry plantations. They also eat grass, fruits, seeds and fungi, and visit fields to graze on clover and eat root vegetables. When frightened a Roe deer gives a gruff bark. The doe calls the kids with a high-pitched piping note.

119 White-tailed deer
Odocoileus virginianus

Identification: Body-length 150–180 cm (58–70 in.); tail fairly long, 15–28 cm (5–11 in.); shoulder-height 90–105 cm (35–40 in.); weight 50–120 kg (110–260 lb). The coat is red-brown in summer, greyish in winter. Rump patch and underside of tail are white (119a). When taking flight the tail is held erect.

Distribution: Southern Canada to the northern Andes; there is an introduced, feral population in south-western Finland.

Habitat and habits: Lives in woods with access to fields and water. This species is not as sociable as the Red deer (114–115), but in other respects its behaviour is much the same. White-tailed deer feed mainly on shoots and leaves of various bushes and trees, although grass and herbs are also eaten.

120 Reindeer
Rangifer tarandus

Identification: Body-length 185–215 cm (72–84 in.); tail about 15 cm ($5\frac{1}{2}$ in.); shoulder-height 105–120 cm (41–47 in.); weight 70–150 kg (154–330 lb) in the male or bull, somewhat less in the cow. Domesticated forms are smaller. Both sexes have antlers, those of the bull being larger. Muzzle hairy. Hooves are broad and can be splayed (120b). Coat long, coarse and thick, grey-brown in summer, paler on neck and flanks, darker on the head, legs and the sides of the belly. Winter coloration paler. The bull has a mane of long hair on the throat. The calves are more brown than grey.

Distribution: From Scandinavia and Finland eastwards through the northern parts of the Soviet Union and Siberia to North America. Introduced into Iceland and Scotland. A smaller race occurs in Spitsbergen. Formerly the upland Reindeer were found along the whole of Scandinavia's mountain chain from the coasts of the Arctic Ocean to southern Norway. About 1800 they were exterminated in the wild state in Sweden, but have survived in the mountainous massif of southern Norway, including the Hardanger area. Another race, the forest Reindeer, is found wild in northern and eastern Finland. Domesticated Reindeer are still kept in considerable numbers by the Lapps in Norway, Sweden and Finland.

Habitat: Upland Reindeer live on mountain heathland all the year, whereas forest Reindeer are found in areas of forest and marsh in summer, and in more open, dry places with a rich growth of lichen during the winter months.

Habits: Upland Reindeer congregate in large herds in winter, sometimes up to 1,000 head, whereas the forest form occurs in smaller herds. Before the calving period in May–June the bulls leave the herds and move about on their own, or in small groups, until mating time in September–October. The cows choose the same place year after year for calving. Each cow has 1 or 2 calves a year. These are able to follow the mother after a few days. The summer herds are formed of the cows with their yearling calves and other young animals. The rutting period is in September–October. The bulls roar and the more powerful succeed in collecting a number of cows. Gestation lasts about 8 months. Summer diet consists of grass and herbaceous plants, as well as the leaves and shoots of various shrubs. In winter the most important food is lichen which they scrape from under the snow with their hooves. In the forests they also eat lichens from the trees as well as fungi. The Reindeer is a migratory species accustomed to wandering over wide areas. The herds walk, trot, gallop and swim well.

121 Elk (Moose in North America)
Alces alces

Identification: Body-length 200–290 cm (78–112 in.); tail very short, 4–5 cm (about 2 in.); shoulder-height 150–220 cm (58–85 in.); weight 320–450 kg (700–990 lb) in the male or bull, 275–375 kg (605–825 lb) in the cow. Summer coat varies from grey-brown to brownish-black; winter coat generally paler. No rump patch. Legs very long and pale. Muzzle large and pendulous. Eyes small, ears large. Mane of dark hair on the throat which is longer in the bull. The hooves are pointed and can be splayed (121c). The antlers of the bull grow out sideways, almost at right angles to the long axis of the body, and there are 2 types: the cervine or branched type (121a) and the palmate or shovel-shaped type (121b). Intermediate types also occur. The antlers, which increase in size each year for the first 10 years of life, are shed in winter and the new antlers are fully formed by August. There may be 12 or more points on each antler. Calves red-brown and without spots.

Distribution: From north-eastern Europe through northern Asia to

northern America. Found throughout Scandinavia except in certain areas of western Norway and along the coasts of the Arctic Ocean. Also throughout Finland, parts of Poland and the forest regions of northern Russia.

Habitat: Open forest with plenty of underplanting. Also in moorland and marshes. In southern Sweden this species has adapted to living in cultivated areas.

Habits: Main activity is early in the morning and towards the evening. Walks and trots, also gallops on occasion and swims remarkably well. During recent years there has been a considerable increase in their numbers in Scandinavia, and in Sweden alone the Elk population has reached over 100,000 individuals. This has presented problems to forestry and agriculture owing to the damage done to crops. Diet consists primarily of the bark and foliage of conifers and deciduous trees—particularly ash, birch and willow—as well as grasses and herbaceous plants. When feeding on aquatic plants they submerge the whole head like a hippopotamus. Cereals and root crops are also eaten. The daily food requirement is about 10 kg (22 lb). Normally solitary but during the winter Elk sometimes live in small herds which are sedentary particularly when there is deep snow. During spring and autumn they may move about quite a lot. The cows calve in May–June, each producing 1 or 2, sometimes 3 calves. A cow with calves is very sedentary, does not associate with other families and will protect her offspring effectively. By the end of August the older bulls have cleaned their new antlers and they become aggressive with the onset of the rutting period. They roar and grunt, attack trees and other objects, and dig large holes in the ground in which they rest and wallow. The cow is only on heat for a few days, usually in September. The period of gestation is about 235 days. After the mating season the young calves follow the mother until the next spring's calving. Sexual maturity is reached by young females at 1½–2 years, by young bulls a year later.

Cattle, sheep and goats

The family Bovidae contains about 115 species distributed in North America, Europe, Asia and Africa, but absent (as wild animals) from Australia. In many species both sexes have horns; these are permanent and not shed annually. The young animal's horns start as a pair of bony knobs on the forehead which soon become fused with the skull. These bony protuberances then grow continuously and form the inner bony core or matrix of the horns; this is covered with a skin which becomes cornified and forms a sheath. New horny layers grow from the base and these form annual rings which are quite distinct. As in the deer, the second and fifth digits are much reduced and may be completely absent. The dentition has 32 teeth, the incisors and canines in the upper jaw being absent. The lower incisors protrude somewhat and press against the gum of the upper jaw as the animal tears off the vegetation on which it feeds.

122 European bison or Wisent
Bison bonasus

Identification: Body-length up to 270 cm (105 in.); tail up to 80 cm (31 in.); shoulder-height 180–195 cm (70–76 in.); weight up to 900 kg (1980 lb) for bulls, 600 kg (1320 lb) for cows. Both sexes have short, curved horns which are set wide apart. Coat long and thick, particularly on the front part of the body which is noticeably taller than the rear part. Coloration includes pale and dark specimens in various shades of red-brown.

Distribution: Formerly found over large areas of central Europe, extending into parts of northern Europe, but eventually it became extinct in the wild. During the 1920s it was re-establishd in one area of the Bialowieza Forest in eastern Poland. The introduced specimens were originally kept in a comparatively small enclosure, but as numbers built up, the herd was allowed to roam farther afield in this extensive forest. At the present time it is doing well. A race that lived in the forests of western

Caucasus was exterminated as late as 1925.

Habitat: Associated with tall primary forest with clearings and dense undergrowth.

Habits: In summer European bison are active throughout most of the 24 hours, except for a few hours in the middle of the day and during the darkest hours of the night. They usually live in herds of 6–30 individuals, but in spring the herds break up to some extent. The older bulls tend to live solitarily while younger bulls may form small groups on their own. The cow lives solitarily during calving, which takes place in May–June; 3–4 days after the birth she joins the herd again with her calf. The cows come on to heat in August–September. The bulls become aggressive and mark the mating area in much the same way as Red deer. Fights occur when the younger bulls attempt to challenge the more powerful males. Growth is slow and the animals are not fully grown until they are at an age of at least 6 years, although sexual maturity is achieved at an age of about 2½ years. Diet consists chiefly of the buds, shoots and foliage of trees; they also eat some grass, herbaceous plants and acorns in the autumn. In winter they strip the bark off trees, and also eat lichens and fungi.

123 Musk-ox
Ovibos moschatus

Identification: Body-length 200–245 cm (78–95 in.); tail short, about 10 cm (3½ in.) and completely hidden in the coat; shoulder-height about 130 cm (50 in.); weight up to 400 kg (880 lb) for the bulls, the cows being considerably smaller and lighter. Both sexes have horns which are broad at the base, meeting in the middle of the forehead where they form a massive boss. Coat dark brown with very thick woolly underfur; the hairs on the belly are up to 70 cm (27 in.) long.

Distribution and habitat: Now found in the wild only in north-east Canada and Greenland, whence it has been introduced into Spitsbergen and Norway. Musk-ox lived in Europe after the Ice Ages but had disappeared long before historic times. They are principally associated with areas of tundra.

Habits: Found in winter in large herds made up of family groups. In spring the individual family groups are re-established, each with a bull, 2 or 3 cows and a number of young animals. In snowstorms, or when threatened by predators, the adults form into a ring with the calves hidden among them or in the centre. The animals stand in this defensive position with heads lowered and facing outwards. This formation provides an effective protection against wolves and other enemies but makes it easier for hunters to shoot the whole herd. Mating takes place in July–September. At this time the bulls fight vigorously for the cows, clashing their armoured foreheads together and even running full tilt at each other from a distance of 20 m (60 ft). In good years a cow may give birth to 2 calves, although usually only a single calf is produced. When feeding is poor she may fail to become pregnant. The gestation period is about 9 months. The calves are able to follow the cows at an age of 2–3 days. Females do not become sexually mature until they are at least 3 years old. Musk-ox feed mainly on the top shoots of arctic willow, but also eat grasses and herbaceous plants. They show great agility in moving across rough, rocky terrain.

124 Mouflon
Ovis musimon

Identification: Body-length 110–130 cm (42–50 in.); tail 3–6 cm (1–2 in.); shoulder-height 65–75 cm (25–30 in.); weight 25–50 kg (55–110 lb). Back red-brown. Pale areas on flanks and on parts of head, legs and belly; rump patch also pale. The male has long, powerful horns which curve backwards and grow to a length of 50–85 cm (20–32 in.). The female sometimes has short horns.

Distribution and habitat: Originally native to the mountain forests of Sardinia and Corsica, but now introduced into many places in southern and particularly central Europe, where they have become well acclimatized and live wild. There are populations of similar sheep on certain islands off Britain; their precise origin is not known.

Habits: Originally associated with mountain and steppe country, including open woodland, but nowadays it is often restricted to regions above the tree-limit. Displays great agility and speed in rocky terrain. Feeds on grasses, sedges and herbaceous plants, also on the leaves and shoots of trees and bushes. It is particularly active during the evening, night and morning. Lives in flocks of 10–20 animals; the males live separately from the ewes except during the rutting period. At lambing time in the spring the ewes leave the flock but rejoin it when their new-born lambs are a few days old. They produce 1–3 lambs at a birth. Prior to mating the rams challenge each other and there is a certain amount of fighting. Mating takes place in late autumn and the gestation period is about 5 months. Mouflon are sexually mature at an age of 1½–2 years. In addition to the usual bleat they make a whistling sound which acts as a warning signal. Like the other forms of *Ovis ammon*, the Mouflon is one of the ancestors of domesticated sheep.

125 Chamois
Rupicapra rupicapra

Identification: Body-length 110–130 cm (42–51 in.); tail 3–4 cm (about 1½ in.); shoulder-height 70–80 cm (27–31 in.); weight up to 50 kg (110 lb) in the male, up to 40 kg (88 lb) in the female. Summer coat yellow-brown with a dark dorsal stripe and paler underparts; long hair in winter, brownish-black with a whitish under-

side (125a). Head pale with a dark lateral stripe. Both sexes have short, slender horns which curve backwards shortly before the tip. There are growth rings on the horns.

Distribution: The Alps, Pyrenees, Carpathians and various mountain ranges in the Balkans, and in Italy and Spain. The distribution was at one time considerably wider.

Habitat: Lives in the highest parts of the forest region from 1,500 m (4,900 ft) to 3,000 m (9,800 ft). In summer often above the tree-limit on the alpine meadows. Prefers steep mountain slopes with deciduous and coniferous trees.

Habits: In winter large herds usually gather in the lower forest regions. If forced to spend the winter above the tree-line, there may be serious losses in a bad winter. In spring, they move up to the tree-line again. The herds break up into smaller groups, the males usually living on their own until the onset of the rutting period in autumn. The kids are born in April–June, normally one in a litter but sometimes more and these are able to follow the mother within a short time.

Both sexes have scent glands and the secretion has a strong odour. These glands are active at the time of the rut. The males start chasing each other and they mark territory by wallowing in pools and rubbing their heads against trees and rocks. They gather a number of females into a harem which they protect vigorously against rivals. Matings take place in late October–December and the gestation period is 6 months. The female is sexually mature at an age of $1\frac{1}{2}$ years. Chamois are particularly active in the morning and afternoon. They are always on the alert and move extremely fast when disturbed. They leap up to 7 m (20 ft) on comparatively flat ground but jump even greater distances when fleeing down a mountainside. Their hooves are narrow and they find adequate footholds on incredibly small surfaces; with their fantastic sense of balance and ability to judge distances, they are remarkably sure-footed even when in full flight.

Each group has a lookout and in case of danger a sharp whistling call acts as a warning. Other sounds include a goat-like bleat. Summer diet includes clover, grasses and various herbaceous plants, as well as the buds and shoots of trees and bushes. In winter they browse on trees and will also eat moss and lichen.

126 Ibex
Capra ibex

Identification: Body-length 130–145 cm (50–56 in.): tail 12–15 cm ($4\frac{1}{2}$–5

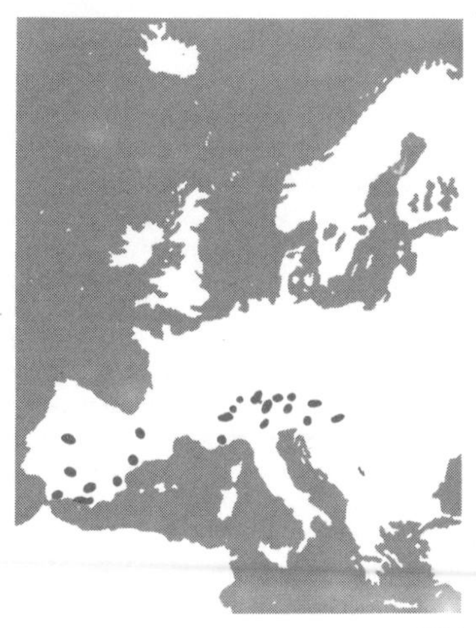

in.); shoulder-height 65–85 cm (25–33 in.); weight up to 120 kg (264 lb) in the male, up to 55 kg (120 lb) in the female. The male's horns curve backwards uniformly and reach a length corresponding to the shoulder-height; those of the female are much shorter. There are growth rings on the horns. Summer coat mainly greyish or brownish-grey, with a black stripe along the belly; paler in winter.

Distribution and habitat: Switzerland, southern Germany, Austria and Italy, in rocky terrain at 2,200–3,200 m (7,200–10,400 ft). Only the Italian population is original. The species was exterminated in the other countries but has been reintroduced from Italy. Other races, smaller than those in the Alps, are found in the Pyrenees, southern Spain and on some of the Greek islands.

Habits: The sexes live apart for much of the year, the males moving about on their own or in small groups, often high up in the mountains in summer. The Ibex lives at higher altitudes than the Chamois and compares favourably with the latter in agility on rocky terrain. Females with kids and juveniles live at somewhat lower altitudes. Activity is greatest in the morning and afternoon. The hours around midday are often spent resting in the same place day after day over a long period. Similarly, well-worn paths to and from their feeding-places are used regularly. Diet consists of grass and herbaceous plants, shoots of shrubs and bushes, and lichen—particularly during winter. Mating takes place in December–January and the kids are born in May–June. Often there is only a single kid in the litter. The gestation is about 6 months. Sexual maturity is reached at an age of 1 year. When alarmed the male produces a sharp whistling sound. The females and their kids bleat rather like domesticated goats.

Primates

The order Primates is divided into 2 main groups: Prosimians which contain exclusively tropical forms, occurring in Madagascar, tropical Africa and south-east Asia, and the Anthropoids which have a large number of species in all parts of the tropics and in a few temperate areas, but are absent from Australia.

It is characteristic of the primates that the skull—and also the brain—are relatively larger than in most other mammals. On the other hand, the muzzle region is usually relatively short. Correlated with this the dentition is somewhat reduced. Each half of each jaw has 2 chisel-shaped incisors, 1 powerful canine—similar in form and size to that of the carnivores—2 or 3 premolars and 3, rarely 2, molars. The cheek teeth (premolars and molars) have 'knobbly' biting surfaces. The large, broad skull makes it possible for the eyes to look forwards, giving binocular vision. Sight is particularly well developed and is one of the most important senses in the primates. The ears are always relatively small and in contrast to the condition in most other mammals—immobile. The limbs are long and slender, quite free from the body along their full length and in most species admirably adapted for climbing. The fore- and hind-feet normally have 5 digits, each with a flat or flattish nail; a few forms have claws on 1 or 2 digits. In some species the first toe may be completely reduced, but in the vast majority it is movable and opposable to the other digits so that the hands and feet are effective grasping organs, adapted both for climbing and for holding food. When eating the food is carried to the mouth by the fore-limbs, unlike most other mammals, which move the head to the food. When walking, the complete sole of the foot is normally placed on the ground. The sound-producing organs are often well-developed and there is a wide repertoire of vocal sounds. Fur is well developed in all the primates, except in the face and buttocks regions; in many species these areas of the body are completely or partly naked and often vividly coloured. The diet is usually vegetarian, either entirely or predominantly, but a considerable amount of animal food is eaten by a number of species, and primates are often described as omnivorous.

The female has 2 nipples on the breast. She usually gives birth to a single young at a time; this youngster clings to her fur and is carried around in this manner for quite a long time. The primates live in social groups; the size of the group varies but the hierarchical structure is nearly always highly organized.

The anthropoids are divided into several families. Two of these occur only in the New World. The Old World monkeys or Cercopithecidae include a large number of species in Africa and Asia, with one species (127) extending into Europe.

The remaining anthropoid families are the Hylobatidae or gibbons, the Pongidae (Chimpanzee, Gorilla, Orang-utan) and the Hominidae (Man).

Cercopithecid monkeys

127 Barbary ape
Macaca sylvana

Identification: Body-length 60–71 cm (24–28 in.); tail very short and hidden in the fur; height at the shoulders about 45 cm (17 in.). The fur has a rough-coated appearance; the colour is brownish or yellowish-grey, paler on the underparts. The fore- and hind-limbs are long and slender and of approximately the same length. The muzzle is relatively short and without fur.

Distribution and habitat: In Europe found only on the cliffs of Gibraltar where the present population lives in semi-wild conditions. Elsewhere it is distributed throughout Morocco and Algeria, where it lives in inaccessible mountainous regions with trees or rocky cliffs.

Habits: These monkeys—the use of the word ape is misleading—live in social groups, each one led by an old male; they are only active in daylight and spend the night in caves in the cliffs. They are omnivorous, feeding on all kinds of animal and plant food. The sound most frequently heard is a sharp yapping but they also have a high-pitched scream. There does not appear to be any particular breeding season. The period of gestation is about 6 months, and the females usually give birth to a single young. The new-born young weighs about $\frac{1}{2}$ kg (about 1 lb) and is looked after by the mother for about a year. Barbary apes become sexually mature at 4 years and have been known to live in captivity for at least 21 years.

Whales

The whales (order Cetacea) number about 90 living species. Whales occur in all the seas of the world and also in certain rivers and large lakes. They are classified in two groups: the toothed whales or Odontoceti in which the jaws have uniform teeth, and the whalebone whales or Mysticeti which lack teeth but have 2 rows of baleen plates hanging down into the mouth from both sides of the upper jaw. The smooth skin has no hair except for a small number of bristles or vibrissae, chiefly in the mouth region. Beneath the thin horny outer skin there is a thicker layer of fat, known as blubber, which provides insulation and food reserves. The nasal channel, ending in a blowhole, is usually situated on the highest point of the head; there are two separate blowholes in the whalebone whales, but in the toothed whales there is only one opening. The outline of the body is elongated and streamlined. Only the fore-limbs are visible and these are reduced to a pair of stiff flippers, movable only at the shoulder joint, which serve as organs of steering and balance. The hind-limbs are reduced to a few bones, deeply embedded in the muscles of the belly. In some whales there is a vertical fold of skin which forms a dorsal fin. The tail is extremely muscular and ends in 2 horizontal tail flukes; these move up and down, and provide the sole means of propulsion. Whales are capable of diving to very great depths and can remain submerged for a long time. The process of breathing in and out takes only a few seconds. When they come to the surface they inhale deeply a few times and also exhale the spent air. This is known as the 'blow' which looks like a jet of water or a spray of foam. The 'blow' may rise to a height of several metres.

Although the ear openings are almost invisible, whales have a well-developed sense of hearing and a large repertoire of sounds. Some of these are audible to the human ear, while others are inaudible like those of the bats. Evidence has now been obtained in some species that high-frequency sounds are used for orientation and for locating prey; doubtless they also play a part in what appears to be a highly complex method of communicating with other members of the species. The eyes are small and sight is far less developed than hearing.

Most species have a gestation period of 11–16 months. As a rule only a single young is born after each pregnancy, but at birth it is already well developed—from a quarter to a third the length of the mother. The milk has a fat content of 30–50 per cent. Instead of being sucked from the nipples, the milk is actively squirted into the young whale's mouth; this process only lasts a few seconds. The young whale (or calf) grows extremely fast.

Five whale families are represented along the coasts of Europe: the dolphins or Delphinidae, the beaked whales or Ziphiidae, the sperm whales or Physeteridae, the rorquals or Balaenopteridae and the right whales or Balaenidae.

Toothed whales

Dolphins

The dolphins in the family Delphinidae include two-thirds of all whale species. Some of these have a clearly defined beak-like snout. Most species are relatively small, and only a few are over 4 m (12 ft) long. The number of teeth varies and is usually large—up to 260. The dolphin family includes some of the fastest and most elegant of all whales. They move around in schools and can jump right out of the water, leaping clear of the surface. Dolphins have a highly organized social life.

128 Porpoise
Phocaena phocoena

Identification: Total length 1·35–1·85 m (4–5½ ft); weight 45–55 kg (92–114 lb). Beak absent, sloping forehead starts immediately from the tip of the jaws. Dorsal fin low and triangular. Back brownish-black to black, belly whitish, but the 2 colours are not sharply demarcated. There are 22–27 teeth in each half-jaw. Unlike other members of the Delphinidae the teeth are not pointed but shaped rather like a spade.

Distribution and habitat: Found along the coasts of Europe from the White Sea in the north to the Mediterranean in the south. Porpoises sometimes swim up the mouths of rivers and enter inland lakes. They are more frequently seen around Britain than any other cetacean species.

Habits: Move about in schools of up to 20 individuals, sometimes more. Undertake regular migrations; for instance, in spring schools move eastwards from Denmark into the Baltic and then back again into the North Sea in autumn. Porpoises breathe about 5 times a minute. When they surface the first part of the body to appear is the back in front of the dorsal fin; this is followed by the dorsal fin itself and finally by the region immediately behind (128a, p. 100). They give the impression of having turned a somersault.

Porpoises feed mainly on fish, especially small cod and whiting; squid are also an important part of their diet. Mating takes place in July–August, and after a gestation period of 10–11 months the female gives birth to a single young, which weighs 6–8 kg (13–17 lb) and is 60–80 cm (23–31 in.) long. This is suckled for about 8 months and becomes sexually mature at an age of about 14 months.

129 Common dolphin
Delphinus delphis

Identification: Total length 1·80–2·60 m (6–8 ft); weight up to 75 kg (165 lb). Body slender, more elegant than preceding species; dorsal fin moderately high and the tip reclines towards the flukes. Back brown or black, belly whitish; undulating stripes of grey or yellow-brown along the sides. Conspicuous beak sharply delineated from the forehead. Each half-jaw has 40–50 small pointed teeth.

Distribution and habitat: Found in warm and temperate seas. Sometimes follows the Gulf Stream northwards and thus reaches the coasts of northern Europe. Occurs more commonly on the coasts of western Europe.

Around the British Isles more often seen off the south and west coasts than in the North Sea.

Habits: One of the fastest whales, speeds of up to 40 km/hr (24 mph) having been recorded. Moves about in schools of from 20 up to several hundred individuals, and often follows ships for long periods. Can jump high out of the water. Feeds on fish, including flying-fish, and squid which are taken near the surface. Becomes sexually mature when body-length is about 1·5 m (5 ft). The female gives birth to a single young in the summer of every second year.

130 Bottle-nosed dolphin
Tursiops truncatus

Identification: Total length of 2·80–4·10 m (9–13 ft); weight 150–200 kg (330–440 lb). Prominent dorsal fin, reclining towards the flukes. Back black or dark grey-brown, belly pale but pigmented from vent to flukes. Receding forehead starts from middle of the beak. Lower jaw protrudes beyond the upper jaw. Each half-jaw has 20–22 teeth.

Distribution and habitat: Found mainly along the coasts of the warm and temperate parts of the Atlantic Ocean and adjacent waters, and is commonest on the east coast of North America. In summer this species reaches as far north as Bear Island and Novaya Zemlya. Occurs off the western coasts of Europe, extending into the Baltic, and also in the whole of the Mediterranean. Also moves up into the lower reaches of rivers. This species is often seen in the English Channel and also occurs off the south and west coasts of the British Isles.

Habits: Not so fast as the preceding species, speeds of up to 30 km/hr (18 mph) having been recorded. Moves about in schools of up to several hundred individuals. Can jump several feet above the surface and is very playful. Feeds on fish. The female mates for the first time at an age of 4 years. The gestation period is about 12 months and a single young is born in alternate years in March–May. It is suckled 2–4 times every hour, throughout the 24-hour period, but on each occasion only for a few seconds, the milk being squirted into its mouth.

131 White-sided dolphin
Lagenorhynchus acutus

Identification: Total length 1·95–2·80 m (6–9 ft). Back black, belly white. Elongated, whitish areas on the side become more yellowish posteriorly and ventrally. Beak short but well defined; upper surface black, not white as in next species.

Each half-jaw has 30–34 teeth. An individual 1·95 m (6 ft) long weighs about 75 kg (165 lb).

Distribution and habitat: Found in the north Atlantic mostly in coastal waters, extending northwards to Greenland and Spitsbergen. Abundant for example, along the Norwegian coast. To the south it reaches the coasts of Belgium and Britain. It is reported more frequently off Orkney and Shetland than in other areas around Britain.

Habits: Often moves about in very large schools, numbering up to 1,000 individuals, and is frequently found in the company of Pilot whales. Feeds on fish, including herring and mackerel, and also on smaller animals. Mating takes place in July–August and the single young is born about 10 months later, in May–June.

132 White-beaked dolphin
Lagenorhynchus albirostris

Identification: Total length 2·35–3·10 m (7½–10 ft). Back black, belly white with greyish areas on the sides. Beak much the same shape as in the preceding species, but upper surface is white. Only 22–25 teeth in each half-jaw.

Distribution and habitat: Found in the northern and north-eastern parts of the Atlantic, occurring most frequently quite near to the coasts. The range extends westwards to Greenland, northwards to Iceland and Tromsø, and southwards to the waters off western France. Off Britain it is more often recorded in the North Sea than on the Atlantic coasts.

Habits: Like the preceding species this dolphin is very gregarious and moves about in large schools. It feeds principally on herring. The single young is born a little later than in the White-sided dolphin, namely July–August, and the next mating takes place some weeks later. Comparatively little is known about the habits of this species.

133 Risso's dolphin
Grampus griseus

Identification: Total length 2·5–4 m (8–13 ft); flippers look narrow and taper to a point, about 60 cm (23 in.) long. Beak absent, forehead rises vertically from front jaws and then recedes slightly at top. Back pale grey to grey-black, according to age, head paler, belly whitish; whitish scar marks are present on the sides of the body. Upper jaw usually toothless, each half of the lower jaw has 2–7 teeth, usually 4.

Distribution and habitat: This species occurs in all oceans, except in the most northerly parts. It is a rare visitor to the coasts of Europe. The most northerly record is from Bohuslän on the west coast of Sweden. Strandings have been reported off Britain, mainly along the south and west coasts.

Habits: Probably lives in schools of less than 12 individuals. Feeds on fish and squid. Little is known about its breeding habits.

134 Killer whale
Orcinus orca

Identification: Total length 3·8–9·5 m (12–30 ft), the male almost twice the length of the female. Prominent dorsal fin in the middle of the body may be over 1 m (3 ft) long in the male. Flippers broad and rounded. Beak absent, forehead slopes down obliquely to the tip of the upper jaw. Back black, belly white but boundary irregular, as white ventral area projects into the black dorsal area towards the rear. Lens-shaped white patch behind the eye. Each half-jaw has 10–30 large teeth that are oval in cross-section.

Distribution and habitat: Found in all seas, including the arctic and subarctic regions. A regular visitor to the coasts of Europe. This species has been recorded off the British Isles on a number of occasions, including stranded specimens.

Habits: Occurs in large or small schools. Swims close to the surface of the sea so that the pointed dorsal fin is often seen sticking up. The normal speed is 10–13 km/hr (6–8 mph), but over short distances much higher speeds are attained. Killers are predators which eat seals and other whales in addition to fish and squid. When seen hunting prey, they give the impression of great ferocity as they 'worry' the prey in much the same way as a pack of hounds bringing down a stag. Bearing in mind that they are large predators, it is only to be expected that they will attack large animals, including the large whalebone whales. Seals are attacked at their breeding-places on the drift ice by killers breaking up the ice on which the seals are lying. In the northern hemisphere mating takes place in the last months of the year, and after a gestation period of about 12 months the female gives birth to a single young which is up to 2 m (6 ft) long.

135 Pilot whale
Globicephala melaena

Identification: Total length 4·3–8·7 m (13–28 ft). A specimen 4·3 m (13 ft) long weighs about 680 kg (1,500 lb). Flippers long and tapering. Dorsal fin about 30 cm. (12 in.) long, rounded at the tip. Almost completely black, with only a narrow, pale greyish stripe on the belly. Forehead bulging, beak inconspicuous. Each half-jaw has 10 teeth which are positioned near the front of the mouth.

Distribution and habitat: Found in the Atlantic Ocean, particularly off Iceland, Faeroes, Orkney and Shetland, and the American coasts extending as far south as New Jersey. Occurs along all the coasts of Europe, and is particularly common off the west coast of Norway. Also found in the Mediterranean and North Sea.

Habits: Undertakes regular migrations between colder and warmer waters. Lives in schools of up to several thousand individuals. The school tends to follow the leader blindly and this sometimes leads to catastrophe, as the whole school may become stranded and die. Thus, in 1874 no fewer than 1,400 individuals died in this way on the east coast of the United States. Pilot whales have been hunted for centuries in the Faeroes, Norway and elsewhere; they are driven into shallow bays where they are killed. Their food consists mainly of squid. The females become sexually mature at an age of 6 years; at this age they are about 3·6 m (12 ft) long. The males do not achieve sexual maturity until they are 13 years old. In the southern and warmer parts of the Atlantic the single young is born in winter. It is 2·5–2·8 m (8–9 ft) long at birth and is looked after by the mother for about 16 months.

136 White whale
Delphinapterus leucas

Identification: Total length 3·6–5·5 m (12–17 ft). Beak inconspicuous, forehead rounded; slight constriction in neck region, dorsal fin absent. Colour varies according to age: dark grey when very young, subsequently becoming mottled brown, then yellowish, and from the age of 4–5 years the species is milky-white. Each half-jaw has 8–10 teeth sloping obliquely forwards.

Distribution and habitat: Occurs in circumpolar arctic waters (Okhotsk Sea, Barents and White Seas,

Greenland and northern Norway). Moves southwards along the west coast of Europe in very bad winters and has even been recorded up the Firth of Forth in Scotland and up the Rhine in Germany, off southern Norway and far into the Baltic Sea. On arctic coasts it keeps to shallow waters.

Habits: Usually moves about in small schools of 5–10 individuals, which may be family groups, but much larger schools are also seen, particularly when on migration. Feeds on squid, fish and crustaceans. Has an audible repertoire of sounds and can be heard over considerable distances. Sailors have referred to this species as the sea canary. The female is sexually mature at an age of 3 years; the gestation period is about 14 months and the female only gives birth every second or third year. The single young is born in March–May. The blubber is relatively thick and the oil extracted from it is of very fine quality. The skin yields a high-quality leather known as 'porpoise-hide'.

137 Narwhal
Monodon monoceros

Identification: Total length 3·9–5·5 m (12½–18 ft). Body pale with numerous dark, rounded markings. Dorsal fin absent but slight ridge on back. Prominent forehead, no beak, but greatly elongated tooth in the male is unmistakable. In both sexes only 2 teeth develop in the upper jaw. In the female these usually remain embedded in the jaw, but in the male the left tooth grows straight out in front of the head; it develops a spiral twist and may grow to a length of over 2·7 m (9 ft). Its function is unknown.

Distribution and habitat: A species ot the high arctic, mostly found in waters north of 70°N, and usually near to the coasts. A very rare visitor along the coasts of north-west Europe, and a few individuals have been recorded as far south as Britain and Holland.

Habits: Usually lives in schools of 6–10 individuals, which are often of the same sex. Large schools of hundreds of individuals have also been seen, probably on migration. A fast swimmer which can dive to a depth of several hundred metres and remain submerged for up to 30 minutes. Food consists of squid, fish and crustaceans. A whistling sound is produced during exhalation at the surface. Little is known of their breeding habits.

Beaked whales

The beaked whale family, Ziphiidae, contains about 14 species which occur in all seas. Reliable information about many of them is lacking, since they live far out at sea, often singly or in small schools. All species have a narrow, beak-like snout. The flippers and dorsal fin are relatively small. There are only 1 or 2 pairs of functional teeth; these are in the lower jaw, towards the front of the mouth and sometimes they project visibly.

138 Cuvier's beaked whale
Ziphius cavirostris

Identification: Total length 5·5–8 m (17–26 ft). Curve of forehead inconspicuous, giving a streamlined appearance. Colour very variable: usually the head and front part of the back are cream-coloured, with the rest of the body black, but some specimens are greyish all over with darker markings on the belly. In the male a pair of large functional teeth—looking like small tusks—project from the tip of the beak in the lower jaw; these teeth are smaller in the female and rarely visible.

Distribution and habitat: Found in all seas, but not in the polar regions. Generally lives far out to sea but a few have been stranded along the Atlantic coasts of western Europe, including the British Isles, and along the Mediterranean coasts in the south. Recorded in the Baltic on several occasions but not in the North Sea.

Habits: This species moves about a lot, usually in schools of 30–40 individuals. It dives for periods of 30 minutes or longer and then remains at the surface for about 10 minutes. Feeds on squid. The male is sexually mature when the body-length is about 5 m (16½ ft), the female when about 6 m (19½ ft). The gestation period is approximately 12 months and at birth the single young is roughly one-third the length of the mother.

139 Sowerby's whale
Mesoplodon bidens

Identification: Total length 4·2–5·6 m (14–18 ft). Top of forehead slightly more pronounced than in the preceding species, the head thus appearing more pointed. Colour mainly blackish above, somewhat paler below. In the male, one tooth in each half of the lower jaw is visible as a flattened and triangular projection, nearly half-way along the lower jaw. In the female the teeth are similar in shape but smaller and usually invisible.

Distribution and habitat: Found in the Atlantic north of the equator, and northwards to Iceland and Norway. Keeps away from coastal waters, although it has been stranded a number of times on the west and south coasts of Europe, including a few specimens off the British Isles and off Norway.

Habits: Lives singly, in pairs or in small schools, and feeds on squid. The skin often shows various scratches and scars, some undoubtedly caused by fights with members of its own species. Mating takes place in the winter or autumn and the young whale is born about 12 months later; it measures about 2 m (6 ft) at birth, is suckled for about 1 year and is weaned when about 3 m (9½ ft) long.

140 Bottlenose whale
Hyperoodon ampullatus

Identification: Total length 7–9·5 m (23–30 ft). The beak is sharply delineated from the forehead which becomes more prominent as the male grows older. The forehead contains an oily substance known as spermaceti. Young individuals are greyish to black, later becoming spotted with yellow, and older individuals are completely yellowish-white. There are 1–2 teeth at the tip of each half of the lower jaw; these are larger in the male than in the female.

Distribution and habitat: Found in winter in the warm parts of the Atlantic Ocean. In the spring it migrates north and spends the summer in the north Atlantic from Greenland in the west to Novaya Zemlya in the east. Often stranded on the coast of western Europe, is reported fairly frequently around the British Isles and has been recorded far into the Baltic Sea.

Habits: Lives in large or small schools, but the older males are often solitary outside the mating season. A very fast swimmer which can jump right up out of the water and dive to a depth of several hundred metres. It normally remains submerged for 10–20 minutes, but harpooned specimens can stay under water for an hour or more. Exhalation produces a 'blow' which is a metre high and can be heard over a long distance. Feeds mainly on squid. The young are born in the spring at an approximate length of 3 m (9½ ft). This species is still hunted to a certain extent in the north Atlantic. A fully-grown male will produce about 2 tons of oil and 100 kg (220 lb) of spermaceti oil.

Sperm whales

The sperm whale family, Physeteridae, contains only 2 species.

141 Sperm whale
Physeter catodon

Identification: Total length normally 15–18 m (48–58 ft) for males, 9–12 m (29–39 ft) for females, but larger individuals have been recorded. A well-grown male may weigh about 50 tons, a female 13 tons. The forehead region is enormously enlarged and extends well beyond the edges of the upper jaw, both forwards and laterally; it may contain up to 5 tons of spermaceti oil. The lower jaw appears disproportionately short and narrow; nevertheless the mouth can be opened so wide that the jaws are nearly at right angles to each other. Each half of the lower jaw has 20–30 teeth; each tooth 15–20 cm (6–7 in.) long. The blowhole is near the front of the head, placed slightly to the left, and is directed obliquely forwards. The dorsal fin takes the form of a low hump and behind it there are a few secondary humps. The flippers are short and rounded. The colour is greyish to almost black, often with a bluish sheen, sometimes paler on the sides and belly.

Distribution and habitat: Mainly found in tropical seas and adjacent waters to about 40° north and south of the equator. In summer, however, males undertake long migrations, dispersing both to the north and south. Some are even found in the Arctic and Antarctic; it is possible that these are unmated males. Stranded individuals have been reported from many parts of western and southern Europe, including the British Isles.

Habits: Usually move about in schools of 15–20 individuals, made up of a male with a harem of females and some young animals. These schools cover large distances in search of food, which consists mainly of large squid, although fish and crustaceans are also eaten. Sperm whales dive to great depths, usually to 300–400 m (980–1,300 ft), but they can 'sound' or go down to about 1,000 m. (3,300 ft). These deep dives normally last for 20–30 minutes, but may extend over more than an hour. When the animal surfaces one first sees the 'hump'. The 'blow' (141a, p. 100) is directed obliquely forwards and may rise to a height of 15 m (48 ft), but each 'blow' only lasts a few seconds. Before sounding again the whale remains at the surface for at least 10 minutes, where it makes a number of quick, shallow dives. The normal speed of a Sperm whale is up to 7 km/hr (4 mph), but when chased it can swim at up to 22 km/hr (13 mph). In spite of its bulk a Sperm whale can jump clear of the water with ease. When starting to sound it goes down almost vertically, with the tail flukes high up above the water surface.

In the northern hemisphere mating takes place in March–May. After a gestation period of about 16 months the female gives birth to a single young, which is about 4 m (13 ft) long.

The teeth have been used as a form of ivory, and the gut sometimes contains large lumps of ambergris which is used in the manufacture of scent. One lump which weighed about 930 lb was valued at about £25,000.

Pygmy sperm whale
Kogia breviceps

Identification: Much smaller than the preceding species, 2·2–4 m (7–13 ft) long, and with a relatively smaller head. Dorsal fin well-defined and reclining. Each half of the lower jaw has 9–14 teeth which are pointed and curve backwards.

Distribution and habitat: Found in tropical seas throughout the world. Stranded individuals have been recorded on the west coasts of France, Portugal and Holland.

Habits: Little is known about this species.

Whalebone whales

Rorquals

The rorqual family, Balaenopteridae contains 5 species which have a worldwide distribution. Like the right whales (family Balaenidae, Nos. 147 and 148), they lack teeth; instead, they have a large number of baleen or whalebone plates hanging down into the mouth from each side of the upper jaw. There are several hundred of these plates which hang one behind the other. They are horny in structure, roughly triangular in shape and 0·2–1·0 m (8–40 in.) long. The inner edge of each baleen plate is frayed to form a fringe which acts as an effective sieve, straining the planktonic animals from the water in the animal's mouth as the large, muscular tongue presses against the sides of the jaw. The body is slender and streamlined with a relatively small head and there are several folds of skin which form grooves or furrows on the ventral surface. The flippers are relatively slender and a dorsal fin is always present. Rorquals are fast swimmers which live mainly on zooplankton, consisting mostly of crustaceans about 1 cm ($\frac{1}{2}$ in.) long which occur in enormous numbers in some parts of the oceans. Rorquals undertake migrations, following their source of food. A few species also eat small, shoaling fish. They breed in warm seas. The females give birth, from their third year, to a single young every year or every alternate year. Rorquals are of great economic importance as the modern whaling industry has largely been built up on these species.

142 Fin whale
Balaenoptera physalus

Identification: Total length 18·5–25m (60–80 ft), the female slightly larger than the male; weight up to 60 tons. The dorsal fin is comparatively small but high, situated far back on the body.

Flippers short and tapering. Forehead meets tip of jaws at an acute angle. There are 70–110 (average about 85) deep grooves, starting at the throat and extending far along the ventral surface. Back grey, belly white. The flippers are pigmented on both surfaces, unlike the next species. The head, however, is asymmetrically coloured: right half of lower jaw white, left half grey, while inside the mouth and on the tongue the coloration is reversed, the right side being pigmented, the left colourless. Seen from above the head is wedge-shaped. There are 320–420 baleen plates in each half of the upper jaw.

Distribution and habitat: Found in all seas, including the Mediterranean. In summer the range extends northwards to arctic and subarctic regions and southwards to correspondingly high latitudes in the southern hemisphere. This species passes the western coasts of Britain on migration and is found in the Atlantic northwards to Spitsbergen and Novaya Zemla, is very common along the coasts of Norway, but more rarely enters the North Sea.

Habits: Usually moves around in large or small schools. The normal speed is about 20 km/hr (12 mph). During the long, deep dives a Fin whale usually remains submerged for 4–15 minutes, but this can probably be extended to about 30 minutes. When surfacing after sounding, it 'blows' 4–5 times in quick succession. Each 'blow' rises vertically, forming a jet 4–6 m (13–19 ft) high (142a, p. 100) which is visible for 3–7 seconds. Normally when diving, the dorsal fin appears at the surface but not the flukes; the whole body, however, is occasionally seen above the surface of the water. Food consists of fish, including herring and other smaller fish, as well as planktonic crustaceans. Mating extends over a period of several months.

Gestation lasts for about 360 days and the calf is suckled for 6 or 7 months after birth. The Fin whale has been one of the most important species in the whaling industry.

143 Blue whale
Balaenoptera musculus

Identification: Total length 22–30 m (72–97 ft), but individuals over 33 m (108 ft) have been recorded. This is the longest and heaviest animal that has ever existed. The weight is usually somewhat over 100 tons. The female is slightly larger than the male. Dorsal fin looks very small, but the flippers look comparatively long and tapering. Seen from above the sides of the head are almost parallel, not wedge-shaped as in preceding species. There are 80–100 (usually about 90) ventral grooves which extend far back on the underside of the body. The colour is blue-grey with paler markings and the undersides of the flippers are white. There are about 360 black baleen plates.

Distribution and habitat: Found in all seas, particularly in the temperate and cold regions, including the north and south Atlantic. In summer this species occurs in small numbers in the waters north of Europe and more rarely to the south along the west coast of Europe. It only rarely enters the North Sea.

Habits: Usually seen in groups of 2 or 3 individuals. Undertakes extensive migrations in search of the best feeding areas, and feeds almost exclusively on the small planktonic crustaceans about 1 cm. ($\frac{1}{2}$ in.) long known as krill. Up to 1,000 kg (2,200 lb) may be found in the stomach. Before sounding or diving deep for a period of 10–20 minutes a Blue whale will make at least a dozen short shallow dives. The 'blow' rises vertically to a height of 7·5 m (8 ft) and is visible for 4–5 seconds. The speed is about 15 km/hr (9 mph) but when pursued this may be more than doubled. The time of mating varies according to latitude but the southern populations mostly pair in June–July when they have moved into warm waters.

After a gestation period of about a year the female gives birth to a single calf which is about 7 m (24 ft) long. During a 24-hour period the mother produces about 500–600 litres (110–130 gallons) of milk which has a fat content of approximately 50 per cent. The young Blue whale grows rapidly and usually becomes sexually mature at an age of 2 years when it is 22–23 m (72–75 ft) long.

144 Lesser rorqual
Balaenoptera acutorostrata

Identification: The smallest of the rorquals. Total length 8–10·5 m (26–33 ft). Dorsal fin relatively high, flippers look very long. Snout comparatively short, triangular when seen from above. About 50 ventral grooves in the region of the throat. Back black or blue-grey, usually sharply demarcated from the white underside. The flippers have a distinctive white patch on the outer surface. Each half of the upper jaw has 260–325 yellowish-white baleen plates which are up to 30 cm (12 in.).

Distribution and habitat: Probably in all seas, dispersing as far as the arctic and antarctic, but most observations are from the north Atlantic. It is the commonest baleen whale off the coasts of Europe, including the British Isles. Extends northwards to Spitsbergen and the White Sea and is common along the Norwegian coast where it goes far up the fjords. Passes through the Skagerrak every summer and enters the Baltic.

Habits: Usually lives singly or in small schools of 3–4 individuals. Each deep dive lasts for 3–7 minutes and is preceded by 5–8 rapid, shallow dives. This whale is often seen playing around ships; it can jump up high above the water and there is at least one description of it leaping out of the water belly uppermost. It feeds partly on fish, partly on planktonic crustaceans. The single young is born after a gestation period of about 10 months. It is about 3 m (9 ft) long at birth, is suckled for about 4 months and becomes sexually mature when about 6·5–7·3 m (21–23 ft) long.

145 Sei whale
Balaenoptera borealis

Identification: Total length up to 18 m (60 ft). More slender than the preceding species. The snout appears long and pointed when seen from above. Dorsal fin relatively conspicuous, flippers comparatively short. There are 32–60, usually about 50 ventral grooves which do not extend very far. Back bluish-black to dark grey, belly white, with an irregular transitional zone. Each half of the upper jaw has about 330 baleen plates which are mainly black, with very fine, soft fringes.

Distribution and habitat: Widely distributed throughout all seas. In winter they live in warmer parts, but in summer they move north or south into colder water, dispersing as far as the ice-edge. In Europe abundant along the coasts of Norway, but now only rarely recorded farther south along the coasts of western Europe. A few strandings are reported from time to time from the British Isles.

Habits: Lives in schools and is probably the fastest of all whales, with an estimated maximum speed of about 48 km/hr (30 mph). The deep dives last for up to 10 minutes. When surfacing the tip of the snout is seen first. The 'blow' is 2–3 m (6–9 ft) high, visible for only a few seconds. The next sounding takes place after a few deep breaths, during which the dorsal fin is clearly seen. The Sei whale feeds principally on planktonic crustacean. The gestation period is about 10–11 months and the single calf is about 4·5 m (15 ft) long at birth. It is sexually mature at an age of $1\frac{1}{2}$ years when it is about 14 m (45 ft) long.

146 Humpback whale
Megaptera novaeangliae

Identification: Total length 11–16 m (36–52 ft). Characterized by the very long flippers which are black or black and white above, but always white underneath. Dorsal fin low and the back with a distinct hump. Tail flukes white below, lack of pigment sometimes spreads to parts of the dorsal surface. This species has a number of knobs and protuberances on the body, particularly on the head and flippers. Barnacles and whale-lice are found on various parts of the body. There are only 14–20 grooves on the underside. Back blackish, belly whitish but often with dark markings. Each half of the upper jaw has 300–320 baleen plates which are greyish-black.

Distribution and habitat: Found in all the oceans of the world, often fairly close inshore, moving into warmer water for breeding and dispersing again into colder water (Arctic or Antarctic) for feeding. No recent records of strandings off the British Isles but this species is known to pass along the western seaboard as specimens used to be caught earlier this century by whalers operating from the Outer Hebrides.

Habits: Living in schools of up to 20 individuals. Swims slowly at a maximum speed of 8 km/hr (5 mph), but appears to indulge in various aquatic acrobatics, often changes course and rolls over, sometimes jumping right out of the water. After a variable number of short dives it will sound for up to 15–20 minutes. When sounding the whole of the tail flukes appear above the surface. The 'blow' (146a, p. 100) is low, making a broad jet, and is accompanied by a loud sound reminiscent of a steamship's siren. Food consists of planktonic crustaceans and small fish. In the northern hemisphere mating takes place in March–April, in the southern in September, and the single young is born after a gestation period of about 11 months. It is about 4·5 m (12 ft) long at birth and is suckled for 5 months until it is about 8 m (25 ft) long. Sexual maturity is achieved at an age of about 2 years when it is some 12 m (39 ft) long.

Right whales

The right whale family, Balaenidae, contains 3 species. These differ from the rorquals in having a more robust build, a relatively larger head, longer baleen plates (1·8–4 m—3½–13 ft) and in having no dorsal fin. The right whales are relatively slow-moving animals which feed exclusively on planktonic crustaceans and small, free-swimming snails which they filter from the water.

147 Black right whale
Eubalaena glacialis

Identification: Total length 14–18 m (45–59 ft). Head proportionately smaller than in the next species, accounting for about a quarter of the total body-length. Strongly arched jaws. There are 220–260 baleen plates in each half of the upper jaw and these are up to 2·5 m (8 ft) long and very flexible. The outer edge of each plate is straight or concave. Body is usually uniform black but there may be irregular patches of white. On the front of the upper jaw there is a yellowish, horny excrescence known as 'the bonnet' which measures roughly 20 × 30 cm (8 × 11 in.). There are smaller excrescences on other parts of the body, and in general this species is subject to infestation by parasites. Also known as the Biscayan or North Atlantic right whale.

Distribution and habitat: Formerly abundant in the Atlantic and other oceans, but this species has been commercially exploited since the Middle Ages and has now become rare as a result of excessive cropping by whalers. At one time the movement in summer took it north of Scandinavia and Finland, but now it is only found in small numbers between 30° and 60°N. There are a few records from the coasts of Europe, including the British Isles, and it has also been seen in the Baltic and Mediterranean Seas.

Habits: Usually lives singly or in pairs, but sometimes in large or small schools. Although faster than the Greenland right whale (148) it only swims at 10–12 km/hr (6–7 mph), hence it was easy prey even for the early whale-hunters. The deep dives last 15–20 minutes and are preceded by 5–6 shallow dives. When sounding the tail appears above the surface. There are two blowholes and the 'blow' (147a, p. 100) is usually double; it is directed forwards and rises to a height of 5 m (16 ft). This whale moves northwards in summer to feed on the rich plankton produced in northern waters. It spends the winter in warmer seas, but not in the tropics. Very little is known about its breeding habits. The species is now protected everywhere and it is possible that in some areas it is beginning to recover in numbers.

148 Greenland right whale
Balaena mysticetus

Identification: Total length 15–21 m (49–69 ft). Head relatively larger than in the preceding species, and up to a third of the total body-length. There are over 300 baleen plates in each half of the upper jaw, each plate being over 3 m (10 ft) in length and very flexible. The outer edges of the plates are convex. Body predominantly black, but white on the chin and front part of the lower jaw is characteristic. Also known as the Bowhead.

Distribution and habitat: A northern form found in the arctic and subarctic regions of the Atlantic and Pacific Oceans, where it lives in areas of drift ice. In the Atlantic sector it has unfortunately become extremely rare, having been almost completely exterminated before the turn of the century by whale-hunters. Formerly this species came south in winter to Iceland, Faeroes and Norway but never moved far from the arctic regions.

Habits: Lives in pairs or small schools, moving along at a speed of only 5–7 km/hr (3–4 mph)—the speed of a man walking. Sounds to a depth of over 1,000 m (3,200 ft) and can remain submerged for up to 1 hour, although usually only for 10–30 minutes. After each sounding it remains at the surface for up to half an hour. The 'blow' is similar to that of the preceding species. Mating takes place in late summer and the gestation period is probably about 1 year. The new-born calf is 3·5–5·5 m (11–18 ft) long and it follows its mother for a year. This species is protected everywhere.

BIBLIOGRAPHY

van den Brink, F. H., *A field guide to mammals of Britain and Europe*, Collins, London, 1967

Brosset, A., *La biologie des Chiroptères*, Masson, Paris, 1966

Corbet, G. B., *The terrestrial mammals of western Europe*, Foulis, London, 1966

Hvass, H., *Mammals of the world*, Methuen, London, 1961

Matthews, L. Harrison, *British mammals*, Collins, London, 1952

Norman, J. R. and Fraser, F. C., *Giant fishes, whales and dolphins*, Putnam, London, 1948

Scheffer, V. B., *Seals, sea lions and walruses. A review of the Pinnipedia*, O.U.P., London, 1958

Simpson, G. G., 'The principles of classification and a classification of mammals', *American Museum Natural History Bulletin* No. 85

Southern, H. N. (ed.), *A handbook of British mammals*, Blackwell, Oxford, 1964

Walker, E., *Mammals of the world* (3 vols), John Hopkins Press, Baltimore, 1964

INDEX

English names

The figures refer to both the illustrations and the descriptions, except that where a page number is given this refers to a mammal that is not illustrated.

Scientific names

The figures refer to both the illustrations and the descriptions, except that where a page number is given this refers to a mammal that is not illustrated.